Audio-Visual
Person Tracking
A Practical Approach

Communications and Signal Processing

Editors: Prof. A. Manikas & Prof. A. G. Constantinides
(Imperial College London, UK)

Communications and Signal Processing – Vol. 4

Audio-Visual Person Tracking
A Practical Approach

Fotios Talantzis

Athens Information Technology, Greece
Imperial College London, UK

Aristodemos Pnevmatikakis

Athens Information Technology, Greece

Anthony G Constantinides

Imperial College London, UK

Imperial College Press

ICP

Published by

Imperial College Press
57 Shelton Street
Covent Garden
London WC2H 9HE

Distributed by

World Scientific Publishing Co. Pte. Ltd.
5 Toh Tuck Link, Singapore 596224
USA office: 27 Warren Street, Suite 401-402, Hackensack, NJ 07601
UK office: 57 Shelton Street, Covent Garden, London WC2H 9HE

British Library Cataloguing-in-Publication Data
A catalogue record for this book is available from the British Library.

Communications and Signal Processing — Vol. 4
AUDIO VISUAL PERSON TRACKING
A Practical Approach

ISBN-13 978-1-84816-581-6
ISBN-10 1-84816-581-1

Printed in Singapore.

To my mother
F. Talantzis

To Efi, Athena and Katerina
A. Pnevmatikakis

To my students, past, present and future
A.G. Constantinides

Preface

Computing systems that are aware of human presence in order to provide heterogeneous services are gaining importance in living and working spaces, in entertainment, security and retail. A central role to such systems is the ability to sense humans and often track them in space across time. Tracking has become a mature topic in radar applications but requires a different set of sensors and algorithms when it involves humans. People generally do not like carrying tracking devices, a fact that inhibits service provision greatly. Instead, person tracking in this book is discussed in one of its unobtrusive flavours i.e. with the use of visual and audio modalities.

This book is about tracking humans using cameras and microphones, focusing on particle filtering algorithms. There are a few excellent texts on tracking [Blackman (1986); Blackman and Popoli (1999)], some of which focus on particle filters [Ristic *et al.* (2004)]. All these texts though focus on radar or sonar tracking. Audio-visual tracking needs different types of measurements on different types of signals: image [Gonzalez and Woods (2007)], video [Tekalp (1995); Forsyth and Ponce (2002); Shapiro and Stockman (2001)] and audio [Brandstein and Ward (2001)] signal processing elements need to be cast into the tracking frameworks. Two early works [Blake and Isard (1998); MacCormick (2002)] paved the way for visual tracking, but audio tracking still lacks a comprehensive text. A recent work covers audio-visual tracking [Zhu and Huang (2007)], mostly from the sensors and applications point of view.

Our aspiration is to fill in the gap between traditional tracking texts and signal processing texts. It is meant to be a solid introduction for the researcher starting in the field but also a good reference for people already working in it. It equips the reader with all the tools to measure the presence of humans in audio and visual signals and convert these measurements into likelihood functions. These likelihood functions are suitable for driving many types of tracking algorithms, but the emphasis is on particle filtering. This became an obvious choice after inspecting the evolution of the relevant literature in the past decade that slowly moved away from deterministic and Kalman versions to the more versatile framework of particle filters.

We believe that the coverage of the material is end-to-end, in the sense that the theoretical foundation of particle filtering and the necessary image, video, audio and array signal processing elements are first established, followed by working examples and MATLAB [Mathworks (2010); Gilat (2004)] implementations. The MATLAB implementations aim to serve as skeletons for the employment of novel systems. We felt that the book would not be complete without a chapter discussing applications and real-world systems. This allowed us to give a more meaningful aspect to an otherwise abstract scientific problem.

This book is written by two generations of authors: Two former students and their PhD supervisor. Thus, in numerous ways the completion of this book would not have been possible without the contribution of the supervisor in terms of guidance, inspiration and patience for over a decade. Additionally, the other two authors hope to have brought into the book a hands-on approach to a rather modern signal processing topic. Either way, throughout the composition, we all felt the obligation to create a book that will familiarise researchers with the topic and quickly enable them to further advance the algorithms.

This book project has not been an easy task. It is the consolidation of years of research in the field, some of it funded by European research projects such as Computers in the Human Interaction Loop [CHIL (2007)], Cognitive Care and Guidance for Active Ageing [HERMES (2010)] and Real-Time Context-Aware and Personalised Media Streaming Environments for Large Scale Broadcasting Applications [e Director 2012 (2010)]. It has been the return-on-investment after the supervision of numerous students, discussions with valuable colleagues and most importantly spending hours away from our families.

October 15, 2010

Fotios Talantzis,

Aristodemos Pnevmatikakis and

Antony G. Constantinides

Acknowledgments

The experience from many tracking systems is depicted in this book. All these systems have been implemented in the Autonomic and Grid Computing Lab of Athens Information Technology with the help of many people, either students (former or current) of the authors, or researchers collaborating with them. The authors wish to thank:

- Andreas Stergiou for the discussions held during endless implementations of visual trackers.
- Nikos Katsarakis for researching and implementing face detectors and particle filter visual and audio-visual trackers.
- Martin and Rasmus Andersen for their MSc theses resulting into a real-time multi-camera and multi-cue particle filter visual 3D tracker. Many thanks also to Prof. Zheng Hua Tan for co-supervising the two theses.
- Panos Papageorgiou for helping with the implementation of our very first visual particle filter tracker.
- Panos Kasianidis and Vasilis Mylonakis for their help in audio direction of arrival estimation systems.
- Ghassan Karame, Genadios Genaro and George Miggos for their implementations of 3D and audio-visual trackers during their MSc theses.
- Elias Rentzeperis, Thodoris Petsatodis and Christos Boukis for sharing their research in voice activity detection.
- Achilleas Anagnostopoulos for his thesis resulting in an augmented reality system with integrated palm tracking.
- Eric Lehmann for his valuable contribution in the early stages of the Audio Tracker.

- Nikos Dimakis and John Soldatos for the middleware design of the AIT SmartLab, that facilitates our live experiments and services.
- The partners in the CHIL, HERMES and My-e-Director 2012 projects (too numerous to list) for valuable discussions and opportunities for testing tracker systems in real application scenarios.
- Nikolaos Konstantinou for proof-reading parts of the manuscript and his suggestions.
- Finally, Nikos Vasilakis for designing the cover of this book.

Contents

List of Figures

List of Tables

Chapter 1

Introduction

1.1 Person tracking

The concept of developing systems that sense people has been popular
for decades. To a large extent this was envisioned by the novels of Orwell [Orwell and Crick (1984)] and Gibson [Gibson (1992)] as ubiquitous
surveillance systems. This is further analysed in [Dey (2001a)] as a set
of technologies "enabling reliable and unavoidable location and tracking of
each individual" in order to support them during their everyday activities.
The authors portray the social implications of such systems (focusing on
potential limitations of personal and social freedoms) in detail. The development of modern sensors, algorithms and computers has brought person
tracking out of the pages of fiction books into our lives. While person
tracking in sparsely populated spaces has been successfully demonstrated
in the past years, monitoring cluttered spaces in order to locate and track
people draws a series of implications for engineers, technology providers,
standardisation boards, and equally importantly for privacy and data pro-
tection commissions. This book focuses on the technologies that can enable
such ubiquitous and accurate person tracking while its effective use relies
on the hands of people designing corresponding applications.

Person tracking is quite different to classical radar tracking, in the sense
that the object to be tracked is within a human habitat. This imposes
constraints on the sensors: the power they emit has to be controlled, they
have to fade into the background, they should operate from a distance
and they should not require special equipment to be carried by the human
targets. These constraints lead to a few possibilities for sensors:

- Pressure sensors: These give a very accurate location, but are
  sparsely positioned, e.g. in relevant pieces of furniture or entrance

mats. One exception is the different versions of touch sensors used for multi-touch devices. These are dense but cover small areas.

- Motion sensors: These give only a presence indication of moving targets in the monitored space. Neither the exact position of the targets, nor their number is known.

- Visual sensors: These are cameras that as a standalone offer localisation on the image plane, but in combination with other cameras can yield 3D positions. Since visual sensors always receive complex signals, it is quite difficult to isolate the targets from the background. This is the clutter problem, that makes visual tracking very challenging.

- Audio sensors: These are microphones and can provide localisation only when used in pairs at the least, by estimating the direction of the arrival of sound. Clutter is a problem here as well. Although most of the background does not generate sound (some appliances do), it reflects sound, causing reverberation.

Cameras and microphones are becoming increasingly cheap, can be easily installed (or are already installed for other applications) and are the only sensors that can provide 3D locations in an unobtrusive manner. Hence audio and visual are the preferred modalities for person tracking. Modern computers and networks give everyone access to the processing power and bandwidth required to replicate, to a large extent, the way humans track people, i.e. using their eyes and ears.

Basic to the problem of localisation and tracking is the type of estimate we are after i.e. the location. By an entity's location we mean a description of its whereabouts, in relation to other known objects or reference coordinates. Additionally, audio-visual location estimation of a person can be ascertained with varying degrees of accuracy depending on the quality of the estimation of the system and the application in hand. Thus, in the scenarios we are examining, location estimates are given with respect to a room corner (indoor applications) or one of the sensors (e.g. an outdoor camera). Alternatively, local coordinates can be converted to a global coordinate system like the Global Positioning System. Given the application, the precision of the localisation systems should be chosen accordingly i.e. a system that can track people with high precision would be unnecessary for an application that tries to detect whether a person is present or absent from a room.

1.2 Why person tracking?

Both the audio and visual signal processing communities have seen large improvements in the technologies they design for human-centric applications. These advances, coupled with declining prices for microphones, cameras and processing systems, now make it practical to deploy sophisticated systems that monitor and react to people and objects in the real world. A series of different organisations as varied as stores, airports, train stations, museums, and homeland security agencies have gradually taken advantage of automated, real-time person-tracking systems for counting people and tracking their movements through indoor and outdoor areas.

The benefits of person tracking technology are especially compelling in four areas:

- Assistive environments: For over a decade, researchers have studied and developed accessible and adaptable environments intended to enhance the autonomy and the quality of life of people. The combination of heterogeneous technologies for such purposes is known as Ubiquitous Computing and the corresponding vision is firmly supported by the Ambient Intelligence concept. In order to be able to offer targeted services to people living in such environments the concept of person tracking remains a central requirement.

- Retail management: By analysing traffic patterns and counting face impressions, retail stores of any type can improve customer service and get access to statistics that determine which displays and products are most effective in terms of their location. It is expected that such traffic analysis will offer retailers a competitive advantage that can make a critical difference to margins and market share.

- Interactive entertainment: Cinemas, museums and contemporary art galleries are exploring new ways to make spaces more interactive. Thus, projected videos and images, lighting, and surround sounds can respond to the location of people. Imagine a museum where exhibits react when they detect the presence of a person in order to swarm a written description around the feet of visitors. A computer system could then propose books and gifts according to the traffic pattern data created by the visitor.

- Security: Probably the most evident application of person tracking lies in the field of surveillance. Person tracking systems can high-

light the location of specific individuals in a large space, such as an
office, a park or an airport. Recent systems use track statistics to
detect suspicious behaviour and trigger an alert.

An analytical review of currently interesting applications and systems
is contained in Chapter 6 where the implications of each application area
are discussed analytically.

1.3   Person tracking challenges

Humans live in environments that generate all sorts of problems for tracking systems. Also, the applications of these systems impose very hard
constraints on the tracking algorithms. The challenges of human tracking
systems are categorised as following:

- Environmental adaptation: A robust person tracking solution
  should be able to operate in cluttered environments where many
  people live and work. In real-world spaces, lighting conditions and
  noise levels can vary greatly, since there is no control over the
  imposed dynamic lighting and noise conditions. This makes it difficult for a system to reliably detect and track individuals. Variable
  weather conditions that alter the amount of lighting arriving from
  a window or simply the operation of an air-conditioning unit that
  adds strongly to the ambient sound field can degrade the quality
  of signals captured by microphones and cameras. Thus, a robust
  tracking system should have an adaptive character in order to ensure operation under any typical circumstances. These problems
  are overcome by suitable processing of the audio and visual systems (see Chapters 3 and 4), as well as by multimodal tracing and
  model adaptation (see Chapter 5).
- Synchronisation: Multi-sensor systems require distributed computer networks where different computers process different audio-
  visual streams. These streams need to be synchronised. This is
  achieved using timestamps and computer clock synchronisation by
  the Network Time Protocol.
- Real-time operation: Depending on the application, tracking people requires real-time analysis and sophisticated algorithms for processing audio-visual data. There is a common misunderstanding
  about what real-time actually means. This is not the rate at which

the sensors provide data, since measurements can be ignored or grouped [Kwok *et al.* (2004)]. The real-time requirements are set by the application. A person tracking system must perform fast enough to provide location estimates at the rate suitable for supporting the actual application and not solely the tracker. For example, if a security system detects somebody fleeing from a vehicle, the system must respond fast enough to raise an alert and interact with other security systems, such as electronic locking systems for doors. Person tracking on its own remains a challenging algorithmic problem but it should be examined in the context of real applications.

- System installation: In typical domestic and office environments the installation of computing infrastructure, cameras and microphones can become impractical. Cabling and cooling requirements must also be considered which often increase cost dramatically. It is thus often the case that the monitoring of an indoor space is based on a few panoramic cameras installed overhead and only a few microphones on tables, instead of the use of multiple cameras and microphones that cover every possible area in the enclosure. Even though the focus of this book lies on the algorithmic part of person tracking such considerations should never be neglected. The problem is alleviated as modern buildings often anticipate installations for smart spaces and the hardware (computers and sensors) constantly shrinks in size.

- Integration and interoperability: Person tracking systems should be able to be integrated with other tracking infrastructure and/or computer applications in need of person tracks. Even though audio-visual tracking systems are still in their infancy in terms of out-of-the-box products, there is a need for standardising their operation so that it becomes compliant with legacy and new infrastructure.

Since all these problems are being addressed by modern technology, algorithms and clever engineering, person tracking is about to become a mature technology for service provision.

1.4   Book organisation

Audio-visual person tracking has gained extensive attention in the signal processing community. The studies have recently converged to algorithms that track people using a consistent Bayesian framework, namely Sequential Monte Carlo (SMC) methods or particle filters (PF). These methods approximate the solution to the person tracking problem using sampling of probability density functions. Their main advantage is that they can deal with the non-linear and non-Gaussian nature of the problem.

In the forthcoming chapters we present PF frameworks that estimate the location of multiple interacting people using measurements from audio and visual streams and propose effective ways to combine the two modalities. While the PF formulation is general, each of the separate problems pose special challenges that are discussed alone prior to their fusion into a full audio-visual system.

This book is organised into six chapters, including the current introductory one. Every chapter begins with an introduction and contains examples with results of the presented algorithms and MATLAB source code for their implementation.

In Chapter 2 we motivate the tracking of humans in terms of their initial detection in space and their subsequent localisation across time and space. The tracking problem is cast into a recursive Bayesian filtering problem, which is then solved by Kalman filters in a special case and particle filters in the general case. The chapter concludes with the introduction of a successful particle filter, namely the Sequential Importance Resampling particle filter (aka Sampling Importance Resampling or Bootstrap particle filter).

Chapter 3 discusses implementations of trackers using solely the audio modality. First we discuss the corresponding system model and the tools that allow us to simulate acoustic environments in order to subsequently optimise the design of real systems. The algorithmic part of the audio tracker is presented by focusing first on the spatial organisation of the microphones and then by introducing different systems for detection and tracking. The chapter also includes a set of simulations in order to quantify the effect of different system parameters.

In Chapter 4 signals from cameras are used to track humans. To that extent, the correspondence between camera views and the real world is first established, followed by a detailed account of what we can measure on still and moving images and how we can integrate these measurements in a visual tracking system. The chapter concludes by detailing end-to-end systems that handle tracking of different types of multiple targets, such as bodies, faces or fingertips.

In Chapter 5 we present a series of mechanisms that fuse different types of measurements. These include different measurements in the same modality (multiple types of visual measurements) or different modalities (audio and video) in order to make the operation of the system more robust under adverse conditions of clutter, illumination and noise.

Finally, in order to demonstrate the importance of person tracking systems in our life, Chapter 6 attempts to review typical applications and corresponding future directions. These affect the way we live, work, entertain and secure ourselves.

This book also has a web page, with valuable information. There the reader can find colour versions of all the figures, the source code discussed throughout the book and the media files used for some of the examples. As the book starts being used, we will also be posting there the unavoidable errata. The web page can be found at:

`http://www.ait.edu.gr/ait_web_site/faculty/apne/AV_tracking_book.html`

Chapter 2

Tracking Algorithms

2.1 Introduction

Tracking of humans involves their initial sensing (termed detection) in space and their subsequent localisation across time and space. For this to be done, humans are represented by a set of features, enumerating measurable properties. The features are measured from signals sensed using different modalities such as visual appearance from cameras, speech captured from microphones, active or passive RFIDs sensed by antennae, body acceleration measured by accelerometers or pressure measured from floor or furniture sensors, to name a few. These are the measurement cues that are used to extract evidence of the existence of humans in a particular region. Different measurement cues are discussed extensively in Chapters 3 and 4 for audio and visual tracking respectively. Their combination is discussed in Chapter 5.

The sensors fall into two categories: those that require humans to carry part of the infrastructure on them (e.g. RFIDs, accelerometers) and those that do not, requiring only infrastructure setup in the monitored space (e.g. cameras, microphones, pressure sensors). In this book we extract features using cameras and microphones, i.e. the equivalent to our eyes and ears, in order not to require wearable devices and to take advantage of the ease of installation of this infrastructure.

In this chapter tracking is introduced, originally deterministic (section 2.3) and then stochastic (section 2.4), leading to recursive Bayesian filtering (section 2.4.1) and two of its solutions, the exact Kalman filtering (section 2.4.2), applicable only under strict conditions, and its numerical approximation in the general case via the particle filter (section 2.4.3). Following from there, the rest of the chapter formulates the particle filter tracker and analyses its options and problems, resulting in the extensively used Sequential Importance Resampling particle filter (section 2.5.6).

2.2 Tracking

A detector is the algorithm that searches for objects of interest in the signals available from the sensors at a particular instance in time. In this book detectors have as targets humans and they operate in the two modalities: audio and visual. Audio detectors provide the location of the source of some human-generated sound, as discussed in Chapter 3. Visual detectors find human skin, faces or whole bodies, as discussed in Chapter 4.

A tracker on the other hand is the algorithm that propagates the knowledge about the location of the object of interest, termed a target, across time and space. Both algorithms suffer from clutter: human presence in audio and visual signals needs to be sensed in the presence of multiple competing irrelevant signals such as sounds from other sources, image patches resembling parts of human bodies or objects occluding part of human bodies. In this challenging environment a tracker needs to utilise all available information from both the audio and visual modalities, as well as from diverse measurement types in these modalities. All this information is measured by different types of detectors.

While it is in principle possible to track some targets by subsequent detections, this is extremely impractical since:

- Detectors are usually computationally expensive.
- Detectors can miss the object when conditions are not favourable. These types of errors are termed misses. They result in gaps in the tracks.
- Detectors can report object existence where there is none. These types of errors are termed false positives and are due to clutter. It is common practice to try to find a balance between the two types of errors, by selecting parameters for the detectors, setting them to a particular operating point. Variations in the operating point occur along the Receiver Operating Characteristic (ROC) curve.
- Temporal information is very useful in fighting clutter. Knowledge of the target location at previous times should assist in localising it in the present.

All the relevant information about the target to be tracked is contained in its state vector $\mathbf{x}$ of dimension $n_{\mathbf{x}}$. In its simplest form the state comprises $n_{\mathbf{x}}$ position variables in the $n_{\mathbf{x}}$-dimensional space. Tracking is sometimes facilitated when more information is provided, forming a model of the target. Simple models to be included in the state can be the size of the target, its orientation, the situation it is in (a person standing or sitting down, the person's identity) and its motion (velocity, acceleration, type of motion). Target models can also be specialised to some sensor type, such as colour models for visual tracking. Hence the state of a target can have a large number of dimensions, with elements that receive continuous, discrete or binary values.

Given the state vector at the previous time instant $\mathbf{x}_{n-1}$ and the sequence of measurement vectors $\mathbf{y}_{1:n}$ since time 1 up to the current time n, the tracker estimates the state at the current time instant $\mathbf{x}_n$. The estimation is usually done in two stages.

First, all knowledge of the motion habits of the target is utilised. This knowledge is encapsulated by the object model. This can be some deterministic kinematic model, such as the constant velocity model, or a probabilistic one introducing uncertainty. The (possibly time-varying) object model $\mathbf{f}_n$ shifts $\mathbf{x}_{n-1}$ into the one-step state prediction $\mathbf{x}_n$ that is based only on the knowledge of the motion habits:

$$\mathbf{x}_n = \mathbf{f}_n\left(\mathbf{x}_{n-1}\right) \tag{2.1}$$

A deterministic object model is given in example 2.1. Probabilistic object models are considered in section 2.5.1.

Example 2.1. The constant velocity model assumes that the velocity of the target does not change.

Consider a one-dimensional tracking problem and a constant velocity object model. Then, given a state $\mathbf{x}_{n-1} = [x, v]^T$, where x is the position and v the velocity at time $n - 1$ and a time step Δt, the current state estimate $\mathbf{x}_n$ is:

$$\mathbf{x}_n = [x + v \cdot \Delta t, v]^T \tag{2.2}$$

■

The second estimation step is the measurement update. The information of the sensors around the one-step state prediction is utilised to update the estimate using some measurements on the sensor information. To measure around the one-step state prediction, a (possibly time-varying) measurement model $\mathbf{h}_n$ is utilised that connects the current state $\mathbf{x}_n$ with the current measurement $\mathbf{y}_n$ of dimension $n_\mathbf{y}$:

$$\mathbf{y}_n = \mathbf{h}_n\left(\mathbf{x}_n\right) \tag{2.3}$$

The measurement model employs a mapping of the state onto a region $R_\mathbf{x}$ of the signal from the sensors (e.g. image patch from a camera) or a property of the signal (e.g. the direction of arrival from a microphone pair). Evidence on the existence of the target is collected from $R_\mathbf{x}$. For person tracking, the region can be on the image plane (two-dimensional - 2D), the floor-plan (2D) or the real-world space (3D), depending on the types and number of available measurement sensors. It spans area in the 2D or volume in the 3D cases. Many different measurement cues can be used during this step. The new state is selected as the best match of the measured cues with a model of the target.

Measurement models and state mappings are discussed throughout the audio and visual tracking chapters.

2.3   Deterministic tracking

Deterministic tracking utilises deterministic object and measurement models. Given the state vector at the previous time instant $\mathbf{x}_{n-1}$ and following the update into the one-step state prediction (based on the object model), the state at the current time instant $\mathbf{x}_n$ can be approximated by searching on a multi-dimensional grid spanning a neighbourhood of the state space centred at the one-step state prediction. The application of the object model to obtain the one-step prediction is depicted in Fig. 2.1.

Every point on the grid corresponds to a candidate state, which in turn can be mapped on some region of the sensor information. In the measurement step, evidence supporting the existence of the target is collected from the region pointed at by each of the candidate states. The measurement step, leading to a refinement of the one-step prediction is also depicted in Fig. 2.1. The grid can be searched extensively, which is guaranteed to yield the optimum state from the candidates, but this can be a computationally intensive process, unsuitable for a real-time implementation. Instead, some efficient grid search algorithm can be used, such as the different search algorithms used for block matching in video compression [Tekalp (1995)].

The success of a tracker based on grid search to find the optimum state $\mathbf{x}_n$ depends on the granularity of the grid and how far it extends around the one-step state prediction. Unfortunately finer grids extending further

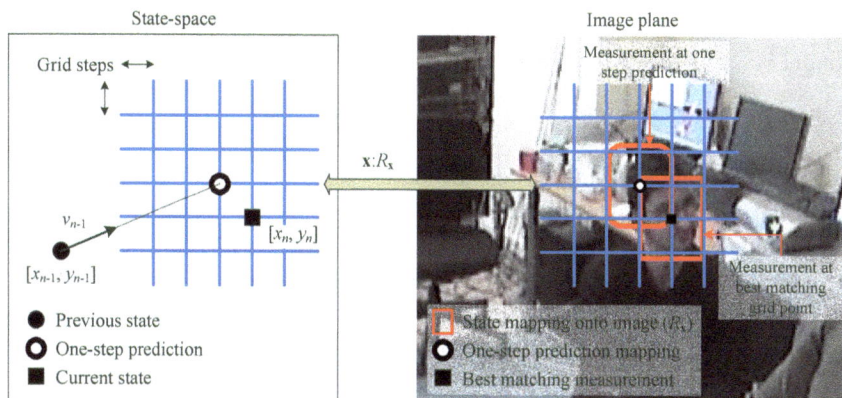

Fig. 2.1 Deterministic tracking as a two-step process. The state at the previous time instance is propagated according to the object model to the one-step prediction. A grid search is conducted around this prediction in the state-space, by mapping all grid points to regions in the signal space (here image plane) and measuring for best match using the measurement model. The best match is mapped back to the state-space, yielding the current state estimation.

away from the one-step state prediction call for many measurement steps, resulting to non-real-time operation.

There are two very successful families of deterministic trackers that are specific to visual tracking, namely variants of template matching (block matching or Lucas–Kanade [Lucas and Kanade (1981)]) and CAM-Shift (Continuously Adaptive Mean-Shift) [Jaffré and Crouzil (2003)]. Since their application is restricted to visual tracking, they are discussed in section 4.10.

2.4 Stochastic trackers

In eqs. (2.1) and (2.3) deterministic object and measurement models have been defined. There are cases though where the relation between the previous state and the current, or the current state and the measurement is stochastic. This happens for two reasons: on the one hand there is noise involved in the measurement process or the target dynamics, or on the other hand there is uncertainty as to their exact nature. The two models then become stochastic or as they are usually termed, dynamic. The dynamic object model is then:

$$\mathbf{x}_n = \mathbf{f}_n \left(\mathbf{x}_{n-1}, \mathbf{v}_n \right) \tag{2.4}$$

where $\mathbf{v}_n$ is a process noise random vector. The dynamic measurement model is:

$$\mathbf{y}_n = \mathbf{h}_n\left(\mathbf{x}_n, \mathbf{u}_n\right) \tag{2.5}$$

where $\mathbf{u}_n$ is a measurement noise random vector.

Due to the stochastic nature of the models, the current state $\mathbf{x}_n$ becomes a random variable. Stochastic tracking, or Bayesian filtering, is then the process of estimating the posterior probability density function (PDF) of $\mathbf{x}_n$, given all the measurement history $\mathbf{y}_{1:n}$, i.e. the conditional PDF of the state, conditioned upon the evidence, $p\left(\mathbf{x}_n | \mathbf{y}_{1:n}\right)$.

2.4.1   *Recursive Bayesian filtering*

Bayesian filtering can be achieved recursively by estimating the posterior $p\left(\mathbf{x}_n | \mathbf{y}_{1:n}\right)$ at time n from the posterior $p\left(\mathbf{x}_{n-1} | \mathbf{y}_{1:n-1}\right)$ at previous time $n-1$ and the current measurement $\mathbf{y}_n$.

Similar to deterministic tracking, recursive Bayesian filtering involves two steps. First the previous posterior is mapped into the one-step prediction density $p\left(\mathbf{x}_n | \mathbf{y}_{1:n-1}\right)$ utilising the object model, i.e. the PDF of the current state $\mathbf{x}_n$ given the previous one $\mathbf{x}_{n-1}$ and the sequence of past measurements $\mathbf{y}_{1:n-1}$:

$$p\left(\mathbf{x}_n | \mathbf{y}_{1:n-1}\right) = \int_{\mathbf{x}_{n-1}} p\left(\mathbf{x}_n | \mathbf{x}_{n-1}, \mathbf{y}_{1:n-1}\right) p\left(\mathbf{x}_{n-1} | \mathbf{y}_{1:n-1}\right) d\mathbf{x}_{n-1} \tag{2.6}$$

where the integration is marginalisation across all possible previous states $\mathbf{x}_{n-1}$.

The posterior is then obtained by utilising the most recent measurement $\mathbf{y}_n$ and Bayes' rule in the measurement update step:

$$p\left(\mathbf{x}_n | \mathbf{y}_{1:n}\right) = \frac{p\left(\mathbf{y}_n | \mathbf{x}_n, \mathbf{y}_{1:n-1}\right) p\left(\mathbf{x}_n | \mathbf{y}_{1:n-1}\right)}{p\left(\mathbf{y}_n | \mathbf{y}_{1:n-1}\right)} \tag{2.7}$$

The denominator $p\left(\mathbf{y}_n | \mathbf{y}_{1:n-1}\right)$ of (2.7) is actually a constant as it depends on the conditioning argument $\mathbf{y}_{1:n}$.

The posterior contains all the information necessary for estimating the target state. According to the Maximum-a-Posteriori (MAP) estimation, the estimated state is the one that maximises the posterior:

$$\mathbf{x}_n^{MAP} = \arg\max_{\mathbf{x}_n}\left(p\left(\mathbf{x}_n | \mathbf{y}_{1:n}\right)\right) \tag{2.8}$$

According to the Minimum Mean Square Error (MMSE) estimate is the expectation of $\mathbf{x}_n$ conditioned upon the evidence, i.e. the measurement history:

$$\mathbf{x}_n^{MMSE} = \int_{\mathbf{x}_n} \mathbf{x}_n p(\mathbf{x}_n | \mathbf{y}_{1:n}) \, d\mathbf{x}_n \tag{2.9}$$

The conditioning upon the past measurement history both in the prediction (2.6) and in the measurement update (2.7) steps make recursive Bayesian filtering impossible. The conditioning on $\mathbf{y}_{1:n-1}$ needs to be eliminated from all terms but the given previous posterior; this is achieved by two simplifying assumptions for conditional independence:

- Given the previous state, the current state is independent of the past measurement history:

$$\mathbf{x}_n \perp \mathbf{y}_{1:n-1} | \mathbf{x}_{n-1} \tag{2.10}$$

  This implies that $p(\mathbf{x}_n | \mathbf{x}_{n-1}, \mathbf{y}_{1:n-1}) = p(\mathbf{x}_n | \mathbf{x}_{n-1})$, which is the object model describing the target dynamics, i.e. the state transition PDF. Hence for the prediction step in (2.6) there is no need for storing the measurement history.

- Given the current state, the current measurement is independent of the past measurement history:

$$\mathbf{y}_n \perp \mathbf{y}_{1:n-1} | \mathbf{x}_n \tag{2.11}$$

  This implies that $p(\mathbf{y}_n | \mathbf{x}_n, \mathbf{y}_{1:n-1}) = p(\mathbf{y}_n | \mathbf{x}_n)$, which is the measurement model. Hence also for the measurement update step in (2.7) there is no need for storing the measurement history. In the form of a PDF, the measurement model is the likelihood of a measurement given the state. The specific forms of the likelihood functions for different types of audio and visual measurements are detailed throughout Chapters 3 and 4.

The recursive form of Bayesian filtering is thus obtained by the one-step prediction step:

$$p(\mathbf{x}_n | \mathbf{y}_{1:n-1}) = \int_{\mathbf{x}_{n-1}} p(\mathbf{x}_n | \mathbf{x}_{n-1}) p(\mathbf{x}_{n-1} | \mathbf{y}_{1:n-1}) \, d\mathbf{x}_{n-1} \tag{2.12}$$

and the measurement update step:

$$p(\mathbf{x}_n | \mathbf{y}_{1:n}) \propto p(\mathbf{y}_n | \mathbf{x}_n) p(\mathbf{x}_n | \mathbf{y}_{1:n-1}) \tag{2.13}$$

The filtering solution of (2.12) and (2.13) is only a conceptual one; marginalising across all possible $\mathbf{x_{n-1}}$ is only analytically tractable in a handful of situations, for example when the state space is finite and discrete (grid-based methods) or as in the Kalman filter discussed in the next section.

2.4.2   *Kalman filter*

When the object and measurement models both have linear deterministic components and Gaussian random components, then there exists an analytic solution to recursive Bayesian filtering, known as the Kalman filter [Kalman (1960); Welch and Bishop (1995)].

The linear dynamic models are defined as follows for the object model:

$$\mathbf{x}_n = \mathbf{F}_n \mathbf{x}_{n-1} + \mathbf{v}_n \tag{2.14}$$

and the measurement model:

$$\mathbf{y}_n = \mathbf{H}_n \mathbf{x}_n + \mathbf{u}_n \tag{2.15}$$

where $\mathbf{F}_n$ and $\mathbf{H}_n$ are matrices of dimensions $n_\mathbf{x} \times n_\mathbf{x}$ and $n_\mathbf{y} \times n_\mathbf{x}$, while the noise vectors $\mathbf{v}_n$ and $\mathbf{u}_n$ are of dimensions $n_\mathbf{x}$ and $n_\mathbf{y}$ respectively.

If the process $\mathbf{v}_n$ and measurement $\mathbf{u}_n$ noises and the initial state $\mathbf{x}_0$ are Gaussian for all n, then the Kalman filter provides the exact recursive Bayesian solution for the posterior as follows.

Gaussian distributions are preserved under linear transformations, hence it is easily shown that both the one-step prediction density and the posterior are Gaussian for all n. As a Gaussian distribution is completely specified by its first two moments, it is sufficient to propagate the conditional mean and the conditional covariance matrix for both the prediction and measurement update steps. Defining the conditional mean as:

$$\mathbf{x}_{n|m} \equiv \mathrm{E}\left[\mathbf{x}_n \,|\, \mathbf{y}_{1:m}\right] \tag{2.16}$$

and the conditional covariance as:

$$\mathbf{C}_{n|m} \equiv \mathrm{E}\left[\left(\mathbf{x}_n - \mathbf{x}_{n|m}\right)\left(\mathbf{x}_n - \mathbf{x}_{n|m}\right)^{\mathrm{T}} |\mathbf{y}_{1:m}\right] \tag{2.17}$$

the mean for the one-step prediction becomes:

$$\mathbf{x}_{n|n-1} = \mathbf{F}_n \mathbf{x}_{n-1|n-1} \tag{2.18}$$

which includes the effect of the deterministic drift dictated by $\mathbf{F}_n$. This drift is illustrated in Fig. 2.2, by the transition from $\mathbf{x}_{n-1|n-1}$ to $\mathbf{x}_{n|n-1}$.

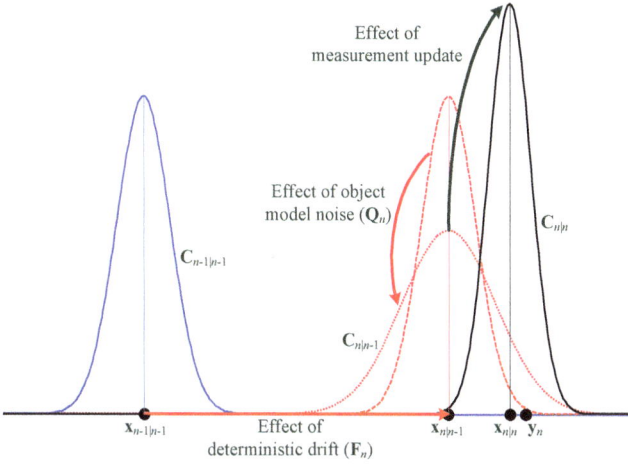

Fig. 2.2 1D example of Kalman filtering. One-step prediction as deterministic drift of the conditional state mean and uncertainty increase, and measurement update by moving the prediction towards the measurement and decreasing the uncertainty based on the Kalman gain. Note that to have the conditional means and the measurement on the same axis, the former needs to be multiplied by the measurement model matrix, which is omitted for simplicity.

The covariance for the one-step prediction becomes:

$$\mathbf{C}_{n|n-1} = \mathbf{F}_n \mathbf{C}_{n-1|n-1} \mathbf{F}_n^{\mathrm{T}} + \mathbf{Q}_n \qquad (2.19)$$

where $\mathbf{Q}_n$ is the covariance of the process noise. What (2.19) implies is that the tracking uncertainty is transformed by the linear component of the object model and is also increased by the object model noise. The increase of the uncertainty from $\mathbf{C}_{n-1|n-1}$ to $\mathbf{C}_{n|n-1}$ is illustrated by the widening of the Gaussian in Fig. 2.2.

For the measurement update, given that $\mathbf{R}_n$ is the covariance of the measurement noise, define the covariance of the innovation term $\mathbf{S}_n$:

$$\mathbf{S}_n \equiv \mathbf{H}_n \mathbf{C}_{n|n-1} \mathbf{H}_n^{\mathrm{T}} + \mathbf{R}_n \qquad (2.20)$$

and the Kalman gain $\mathbf{K}_n$:

$$\mathbf{K}_n \equiv \mathbf{C}_{n|n-1} \mathbf{H}_n^{\mathrm{T}} \mathbf{S}_n^{-1} \qquad (2.21)$$

Obviously larger measurement noise covariances decrease the Kalman gain. The mean for the measurement update then is:

$$\mathbf{x}_{n|n} = \mathbf{x}_{n|n-1} + \mathbf{K}_n \left(\mathbf{y}_n - \mathbf{H}_n \mathbf{x}_{n|n-1} \right) \qquad (2.22)$$

indicating that the difference between the measurement and the prediction is bridged to an extent proportional to the Kalman gain. Hence smaller measurement noise covariances result in the measurement update step driving the state closer to the measurement than to the prediction.

The covariance for the measurement update is:

$$\mathbf{C}_{n|n} = \mathbf{C}_{n|n-1} - \mathbf{K}_n \mathbf{H}_n \mathbf{C}_{n|n-1} \qquad (2.23)$$

indicating that the track uncertainty decreases by a factor proportional to the Kalman gain. The measurement update step of the Kalman filter is also illustrated in Fig. 2.2, by the final state mean moving from the prediction towards the measurement and the uncertainty after the one-step prediction decreasing, as indicated by the narrower Gaussian.

The design of the dynamic models and the Kalman filter is demonstrated in example 2.2.

Example 2.2. Kalman filter with 2D position and constant velocity object model.

Assume a target moving along a curved 2D path. The actual path is unknown, but it is measured, albeit with some noise. A Kalman filter estimates the path employing a constant velocity object model, with the uncertainty regarding its validity being expressed by Gaussian process noise, and a measurement model with Gaussian noise representing the measurement error.

First a curved 2D path is constructed in MATLAB. It comprises N points in a four-dimensional state-space, where two are the position and another two are the velocity dimensions. The time step is Dt:

```
Dt=1/15;
N=20;
k=1:N;
v=[sin(1+k/(.15*N));.5*cos(2+k/(.17*N))];
v=v*15000/N;
x=zeros(4,length(k));
x(:,1)=[170;v(1,1);200;v(2,1)];
for k=2:N
    x([1 3],k)=x([1 3],k-1)+x(:,k)*Dt;
    x([2 4],k)=v(:,k);
end
```

The actual states x are hidden. What is observed is a set of noisy position only measurements y, which are Gaussian perturbations of the states, generated in MATLAB as follows:

```
s2=40;
y=x([1 3],:)+sqrt(s2).*randn(size(x([1 3],:)));
```

The square root of variance is the expected RMS (Root-Mean-Square) error of the measurements from the hidden states. This is 6.32.

The design of the Kalman filter begins with the dynamic models, whose assumptions and MATLAB implementations are discussed next.

The linear part $\mathbf{H}_n$ of the measurement model in (2.15) is constant across time and is simply an association of the two position elements of the four-dimensional state to the two elements of the measurement vector:

$$\mathbf{H} = \begin{bmatrix} 1 & 0 & 0 & 0 \\ 0 & 0 & 1 & 0 \end{bmatrix} \qquad (2.24)$$

The measurement model noise $\mathbf{R}_n$ in (2.15) is also constant across time. Since no knowledge is available that relates the horizontal and vertical positions, they are assumed independent. The measurement noise covariance matrix then is:

$$\mathbf{R} = \sigma_R^2 \mathbf{I}_2 \qquad (2.25)$$

where $\mathbf{I}_2$ is the 2×2 identity matrix and the variance of each element position element is σ_R^2.

The measurement model is initialised in MATLAB as follows, with σ_R^2 denoted by s2R:

```
H=[1 0 0 0;...
   0 0 1 0];
R=s2R*eye(2);
```

The object model is the constant velocity model as the one discussed in example 2.1 only extended for two dimensional motion. Hence the linear term $\mathbf{F}_n$ of the dynamic model in (2.14) is again constant across time and is given by the matrix $\mathbf{F}$:

$$\mathbf{F} = \begin{bmatrix} \mathbf{F}_{cv} & \mathbf{0}_2 \\ \mathbf{0}_2 & \mathbf{F}_{cv} \end{bmatrix} \qquad (2.26)$$

where $\mathbf{0}_2$ is the 2×2 zero matrix, $\mathbf{F}_{cv} = \begin{bmatrix} 1 & \Delta t \\ 0 & 1 \end{bmatrix}$ governs the update of any of the two position elements given the velocity ones, and Δt is the inverse of the frame rate. The covariance matrix of the process noise is:

$$\mathbf{Q} = \begin{bmatrix} \mathbf{Q}_{cv} & \mathbf{0}_2 \\ \mathbf{0}_2 & \mathbf{Q}_{cv} \end{bmatrix} \qquad (2.27)$$

where $\mathbf{Q}_{cv} = \begin{bmatrix} \Delta t^3/3 & \Delta t^2/2 \\ \Delta t^2/2 & \Delta t \end{bmatrix}$ is the 2×2 covariance term corresponding to any of the two position–velocity pairs according to the constant velocity model [Blackman and Popoli (1999)]. It is this process noise that allows the velocity of the target to change no matter the constant velocity object model.

The object model is initialised in MATLAB as follows, with σ_R^2 denoted by s2R:

```
F=[1 Dt 0 0;...
   0 1  0 0;...
   0 0  1 Dt;...
   0 0  0 1];
Q1=[Dt^3/3 Dt^2/2;Dt^2/2 Dt];
Q=[Q1 zeros(2);zeros(2) Q1];
```

The final initialisation step is for the arrays holding the estimates of the state mean (xkk) and covariance (Ckk), the one-step predictions of those (x01 and C01), covariance of the innovation term (S) and the Kalman gain (K). It is not needed to keep all this history, but it is convenient for generating the plots further down the example. Finally the initial state estimate is set to the first measurement for position and the difference between the first two measurements divided by the time step for velocity:

```
xkk=zeros(4,N);
Ckk=zeros(4,4,N);
x01=zeros(4,N);
C01=zeros(4,4,N);
S=zeros(2,2,N);
K=zeros(4,2,N);
xkk(:,1)=[y(1,1);(y(1,2)-y(1,1))/Dt;y(2,1);(y(2,2)-y(2,1))/Dt];
```

The Kalman filtering itself is done as follows:

```
for k=2:N
    % Prediction
    x01(:,k)=F*xkk(:,k-1);
    C01(:,:,k)=F*Ckk(:,:,k-1)*F'+Q;
    % Covariance of inovation term
    S(:,:,k)=H*C01(:,:,k)*H'+R;
    % Kalman gain
    K(:,:,k)=C01(:,:,k)*H'*inv(S(:,:,k));
    % Measurement update
    xkk(:,k)=x01(:,k)+K(:,:,k)*(y(:,k)-H*x01(:,k));
    Ckk(:,:,k)=C01(:,:,k)-K(:,:,k)*H*C01(:,:,k);
end
```

The ratio of the measurement noise variance σ_R^2 over the variance of the process noise (say the position element, $\Delta t^3/3$) determines the relative trust one puts on the measurements. Very small ratios produce paths that blindly follow the noisy measurements, ignoring the object model. Very large ratios produce paths that follow just the deterministic part of the object model (in this example motion along a line with constant velocity), ignoring the measurements. Intermediate ratios are the useful ones, resulting in compromises between the deterministic part of the object model and the measurements. Paths for these three cases are shown in Fig. 2.3. For a ratio of 0.25, the measurements are followed very closely, resulting in an RMS error between the state estimates and the actual hidden states of 5.79. In this case trusting the measurements can not hurt too much; the measurements are not very noisy. For a ratio of 2, trust is better balanced between the object model and the measurements, resulting in a smaller RMS error of 4.75. Finally, for a ratio of 64, the object model is followed more closely than it should (since obviously for this example the velocity is far from constant), resulting in a track that resists turning at the curved portions of the path and a large RMS error of 22.1. The RMS error as a function of the ratio of measurement over the process noise variances is shown in Fig. 2.4.

Note that the optimum ratio measurement and process noise variances depend on the quality of the two models in terms of how closely they match the hidden state process and on the quality of the measurement system, i.e. the sensors themselves and the signal processing that extracts the measurements from the signals recorded by the sensors. Many cases of such signal processing algorithms are detailed in Chapters 3 (for audio) and 4 (for moving images).

■

Due to its closed form solution and ease of implementation, the Kalman filter is perhaps the most widely used form of recursive Bayesian filtering and has been extensively used in tracking applications [Welch and Bishop (1995); Jang *et al.* (2002); Weng *et al.* (2006); Pnevmatikakis and Poly-menakos (2006); Karame *et al.* (2007)]. An application for finger tracking is detailed in section 4.11.1.1. It has though two serious shortcomings that necessitate other approaches:

- The dynamic models need to be linear. This is usually overcome by local linearisation at the current state, resulting in the extended Kalman filter [Orderud (2005)].

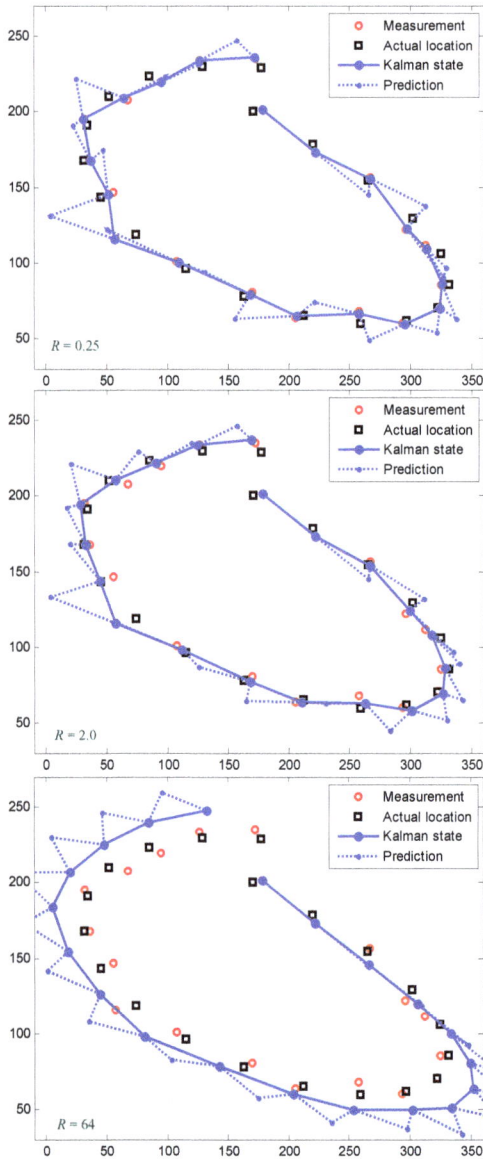

Fig. 2.3   2D tracks from example 2.2, for three values of the ratio of the measurement over the process noise variances. Only the position is shown for the hidden states (actual locations), the state estimates and the one-step predictions. For the latter, dotted lines begin from the previous state estimate and finish at the predictions. A second dotted line connects the predictions with the current estimate based on the measurement update.

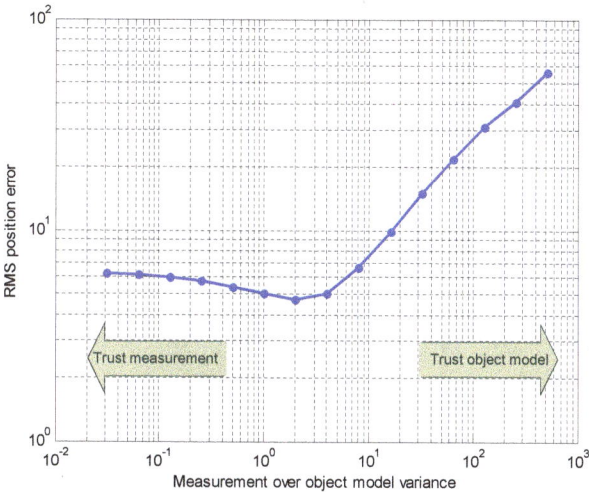

Fig. 2.4 RMS error between the estimated and the hidden states as a function of the ratio of the measurement over the process noise variances for example 2.2.

- The posterior is uni–modal Gaussian. This is the most serious shortcoming of the Kalman filter and is addressed by the particle filter.

2.4.3 *Particle filter*

The particle filter [Arulampalam *et al.* (2002); Pérez *et al.* (2004); Isard and Blake (1998)] is a Monte–Carlo solution for recursive Bayesian filtering. It offers two important advantages:

- It provides a solution in the case of generic object and measurement dynamical models. There is no need for the linearity and most importantly the Gaussianity constraints.
- It provides an elegant means of fusing multiple measurement cues from different modalities, by means of partitioned sampling. This will be investigated further in Chapter 5.

On the other hand, as any numerical approach yielding no closed-form solution, the particle filter has two disadvantages:

- It provides only an approximate solution.
- It is computationally intensive.

Particle filtering is based on two principles: the approximation of distributions with discrete particles (see section 2.4.3.1) and importance sampling (see section 2.4.3.2).

2.4.3.1 *Approximating distributions using particles*

A target distribution $p(\mathbf{x})$ can be approximated by a set of N_p discrete samples $\mathbf{x}^{(i)}$ and their associated weights $w^{(i)}$. Jointly the samples and the weights form the particles, $\left\{\mathbf{x}^{(i)}, w^{(i)}\right\}_{i=1}^{N_p}$. The approximation then is:

$$p(\mathbf{x}) \simeq \sum_{i=1}^{N_p} w^{(i)} \cdot \delta\left(\mathbf{x} - \mathbf{x}^{(i)}\right) \tag{2.28}$$

where $\delta(\bullet)$ is the Kronecker delta function.

A one-dimensional bimodal distribution comprising two Gaussians and its approximation by 20 particles is shown in Fig. 2.5. The particles are illustrated as disks scattered in the state-space, whose centres represent the sample values $\mathbf{x}^{(i)}$ of the particles and whose radii are proportional to the weights $w^{(i)}$ of the particles.

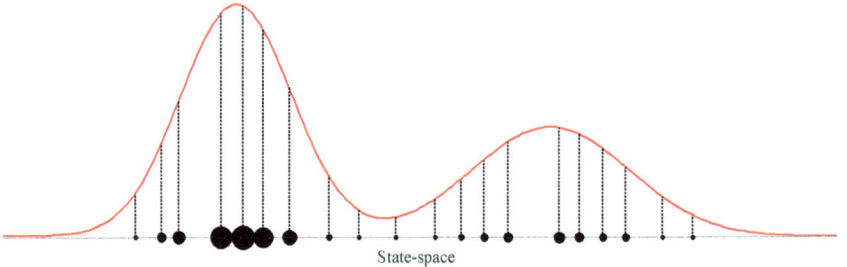

State-space

Fig. 2.5 Representation of a distribution with particles. The particles are depicted as disks, with centres corresponding to the sample values and radii proportional to the weights of the particles.

The particles can then be used to estimate any moment of the distribution. For example, the expected value of $\mathbf{x}$ is:

$$E[\mathbf{x}] = \sum_{i=1}^{N_p} \mathbf{x}^{(i)} p(\mathbf{x}^{(i)}) \tag{2.29}$$

Then replacing (2.28) into (2.29) yields:

$$E[\mathbf{x}] \simeq \sum_{i=1}^{N_p} \mathbf{x}^{(i)} \sum_{k=1}^{N_p} w^{(k)} \cdot \delta\left(\mathbf{x}^{(i)} - \mathbf{x}^{(k)}\right) = \sum_{i=1}^{N_p} w^{(i)} \mathbf{x}^{(i)} \tag{2.30}$$

2.4.3.2 *Importance sampling*

There are cases where the target distribution $p(\mathbf{x})$ can be evaluated for any $\mathbf{x}$, but samples cannot be drawn from it. Instead they can be drawn from a proposal or importance distribution $q(\mathbf{x})$, yielding the set $\left\{\mathbf{x}^{(i)}, 1/N_p\right\}_{i=1}^{N_p}$. According to importance sampling theory, for the particles drawn from the proposal distribution to approximate the target one, the weights are calculated as:

$$w^{(i)} \propto \frac{p(\mathbf{x}^{(i)})}{q(\mathbf{x}^{(i)})} \qquad (2.31)$$

Note that the weights in (2.31) have to be normalised to sum to unity.

Importance sampling for the bimodal distribution of Fig. 2.5 is demonstrated in example 2.3.

Example 2.3. Importance sampling of a bimodal distribution.

The target distribution in Fig. 2.5 comprises two Gaussians:

$$p(x) = \beta N\left(x \,|\, m_1, C_1\right) + (1 - \beta)N\left(x \,|\, m_2, C_2\right) \qquad (2.32)$$

with means, variances and relative weight given by the MATLAB variables m1=3, m2=7, C1=.5, C2=1 and b=.6.

Even though drawing samples from such bimodal distributions is easy (see example 2.5), for the sake of this example importance sampling is performed. The chosen proposal distribution is a much wider Gaussian, with mean m given by the weighted average of the two means and variance C covering twice as much range as the two Gaussians of the target:

```
m=b*m1+(1-b)*m2;
C=((m2+3*sqrt(C2)-(m1-3*sqrt(C1)))/3)^2;
```

Np=20 samples are drawn from the Gaussian proposal using **randn** and are stored in vector **x**:

```
x=randn(1,Np)*sqrt(C)+m;
```

The implementation of the weight determination eq. (2.31) involves evaluating the target and proposal distributions at the drawn samples **x**:

```
% Evaluate proposal at particles
qx=normpdf(x,m,sqrt(C));
% Evaluate target at particles
px=b*normpdf(x,m1,sqrt(C1))+(1-b)*normpdf(x,m2,sqrt(C2));
% Importance weights
w=px./qx;
w=w/sum(w);
```

The target and proposal distributions, as well as the 20 drawn particles are shown in Fig. 2.6.

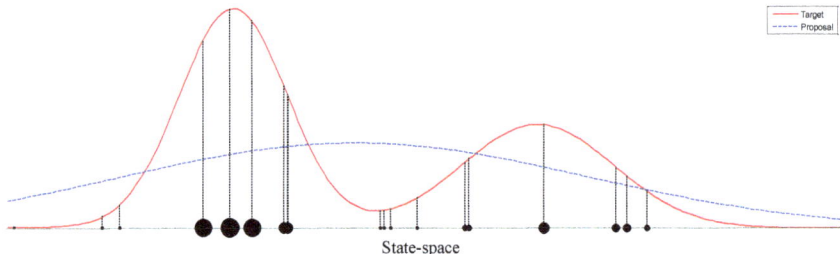

Fig. 2.6 Importance sampling of a bimodal target distribution using a wide Gaussian proposal.

■

Although the use of the weights given by (2.31) in theory facilitates importance sampling, in practice the number of particles for an efficient representation of $p(\mathbf{x})$ depends on how representative of the target distribution is the proposal distribution. If the chosen $q(\mathbf{x})$ is not representative of $p(\mathbf{x})$, then the necessary N_p can be large, incurring severe processing penalties. Example 2.4 discusses the effect of different proposal distribution choices for a triangular target distribution.

Example 2.4. Sample the triangular one-dimensional distribution with a peak at 0.5, extending from -2.5 to 3.5, shown in Fig. 2.7.

Since it is not straightforward to draw samples from the triangular distribution, three different proposal distributions are used. One is uniform and samples are drawn using the **rand()** MATLAB function. The other two are Gaussians, one narrow as to include the support of the target PDF and another much wider. The Gaussian samples are drawn using **randn()**. All three proposals are zero-mean, the uniform extends from -5 to 5, the narrow Gaussian has a standard deviation of 1.5 and the wide Gaussian has one of 4.5. The proposal PDFs are also shown in Fig. 2.7.

In order to obtain the weights according to (2.30), the target and proposal distributions need to be evaluated at the values of the samples $x^{(i)}$, $i = 1, \ldots, N_p$. This is done using the MATLAB function **normpdf()** for the Gaussian proposals and straightforward evaluation of the piecewise linear triangular and uniform distributions.

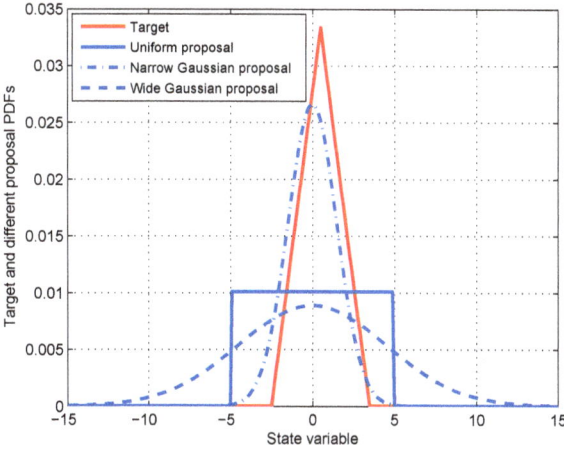

Fig. 2.7   Target and proposal distributions for the importance sampling of example 2.4.

In order to evaluate the efficiency of each of the proposal distributions, the resulting particles are used to estimate the mean of the target distribution according to (2.28). The quality of the mean estimations depends on the effectiveness of the chosen proposal distribution and the number of particles in the representation. Since drawing samples from a distribution is a random process, 2,000 draws of N_p samples are made. For each set of samples the estimation error of the mean of the target distribution is calculated. The root-mean-square of this error (RMSE) is found across the 2,000 draws. The dependency on N_p and the choice of proposal distribution is shown in Fig. 2.8.

Fig. 2.8 demonstrates the expected result that increasing the number of particles improves the estimation of the target distribution. It also shows that proposal distributions matching the target distribution more closely are more efficient: to obtain the RMSE performance of 50 particles drawn from the narrow Gaussian proposal, one needs 68 particles from the uniform one and 76 from the wide Gaussian.

∎

2.5   Particle filter trackers

According to the particle filtering framework, a set of N_p particles $\left\{ \mathbf{x}_{n-1}^{(i)}, w_{n-1}^{(i)} \right\}_{i=1}^{N_p}$ approximating the posterior $p\left(\mathbf{x}_{n-1} \mid \mathbf{y}_{n-1}\right)$ at time $n-1$

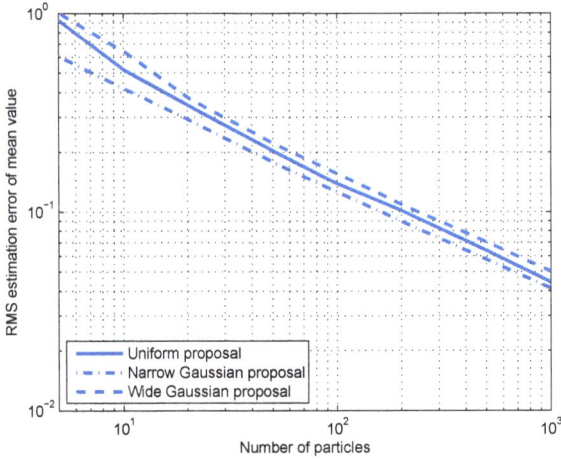

Fig. 2.8   RMSE of the estimation of the mean value of the target distribution for the three different proposal distributions of example 2.4 (see Fig. 2.7) as a function of the number N_p of particles used.

is updated to $\left\{ \mathbf{x}_n^{(i)}, w_n^{(i)} \right\}_{i=1}^{N_p}$ approximating the posterior $p\left(\mathbf{x}_n \mid \mathbf{y}_n\right)$ at time n.

To develop the particle filter recursion, one considers importance sampling for a target distribution $p\left(\mathbf{x}_{0:n} \mid \mathbf{y}_{1:n}\right)$ by means of a proposal distribution $q\left(\mathbf{x}_{0:n} \mid \mathbf{y}_{1:n}\right)$. According to importance sampling and (2.31), the weights at time n are given by:

$$w_n^{(i)} \propto \frac{p\left(\mathbf{x}_{0:n}^{(i)} \mid \mathbf{y}_{1:n}\right)}{q\left(\mathbf{x}_{0:n}^{(i)} \mid \mathbf{y}_{1:n}\right)} \tag{2.33}$$

Similarly, the weights at time $n-1$ are:

$$w_{n-1}^{(i)} \propto \frac{p\left(\mathbf{x}_{0:n}^{(i)} \mid \mathbf{y}_{1:n-1}\right)}{q\left(\mathbf{x}_{0:n}^{(i)} \mid \mathbf{y}_{1:n-1}\right)} \tag{2.34}$$

Since the proposal distribution is selected to be convenient (but also representative of the target one) it can be chosen to factorise as:

$$q\left(\mathbf{x}_{0:n} \mid \mathbf{y}_{1:n}\right) = q\left(\mathbf{x}_n \mid \mathbf{x}_{0:n-1}, \mathbf{y}_{1:n}\right) q\left(\mathbf{x}_{0:n-1} \mid \mathbf{y}_{1:n-1}\right) \tag{2.35}$$

If the target distribution can also be factorised as:

$$p\left(\mathbf{x}_{0:n} \mid \mathbf{y}_{1:n}\right) = A\left(\mathbf{x}_{0:n}, \mathbf{y}_{1:n}\right) p\left(\mathbf{x}_{0:n-1} \mid \mathbf{y}_{1:n-1}\right) \tag{2.36}$$

where $A\left(\mathbf{x}_{0:n}, \mathbf{y}_{1:n}\right)$ is an arbitrary fuction of the state and measurement history, then, replacing (2.35) and (2.36) into (2.33) and utilising (2.34), the weights can be recursively calculated as:

$$w_n^{(i)} \propto w_{n-1}^{(i)} \frac{A\left(\mathbf{x}_{0:n}^{(i)}, \mathbf{y}_{1:n}\right)}{q\left(\mathbf{x}_n^{(i)} \middle| \mathbf{x}_{0:n-1}^{(i)}, \mathbf{y}_{1:n}\right)} \qquad (2.37)$$

The factor $A\left(\mathbf{x}_{0:n}, \mathbf{y}_{1:n}\right)$ in (2.36) is found by first applying Bayes' rule on the target distribution:

$$p\left(\mathbf{x}_{0:n} \middle| \mathbf{y}_{1:n}\right) = \frac{p\left(\mathbf{y}_n \middle| \mathbf{x}_{1:n}, \mathbf{y}_{1:n-1}\right) p\left(\mathbf{x}_{0:n} \middle| \mathbf{y}_{1:n-1}\right)}{p\left(\mathbf{y}_n \middle| \mathbf{y}_{1:n-1}\right)} \qquad (2.38)$$

As already discussed for (2.7), the denominator is a constant and can be omitted since the factorisation is only needed up to a constant. This is because the weights are calculated only up to a constant of proportionality and are subsequently normalised to unity sum. Also, from the definition of conditional probability:

$$p\left(\mathbf{x}_{0:n} \middle| \mathbf{y}_{1:n-1}\right) = p\left(\mathbf{x}_n \middle| \mathbf{x}_{0:n-1}, \mathbf{y}_{1:n-1}\right) p\left(\mathbf{x}_{0:n-1} \middle| \mathbf{y}_{1:n-1}\right) \qquad (2.39)$$

Using (2.38) and (2.39), the necessary factorisation of the target distribution is achieved with:

$$A\left(\mathbf{x}_{0:n}^{(i)}, \mathbf{y}_{1:n}\right) \equiv p\left(\mathbf{y}_n \middle| \mathbf{x}_{1:n}, \mathbf{y}_{1:n-1}\right) p\left(\mathbf{x}_n \middle| \mathbf{x}_{0:n-1}, \mathbf{y}_{1:n-1}\right) \qquad (2.40)$$

Now, given the conditional independence assumption of (2.10), it is:

$$p\left(\mathbf{y}_n \middle| \mathbf{x}_{1:n}, \mathbf{y}_{1:n-1}\right) = p\left(\mathbf{y}_n \middle| \mathbf{x}_{1:n}\right) \qquad (2.41)$$

Also, given the conditional independence assumption of (2.11), it is:

$$p\left(\mathbf{x}_n \middle| \mathbf{x}_{0:n-1}, \mathbf{y}_{1:n-1}\right) = p\left(\mathbf{x}_n \middle| \mathbf{x}_{0:n-1}\right) \qquad (2.42)$$

Two more simplifying assumptions are due:

- The state sequence is Markov, i.e. only the previous state is consequential for estimating the current one, not the rest of the state history:

$$\mathbf{x}_n \perp \mathbf{x}_{0:n-2} \middle| \mathbf{x}_{n-1} \qquad (2.43)$$

- The measurements are conducted through a memoryless channel, i.e. the current measurement, conditioned upon the current state, is independent of the rest of the state sequence and all other measurements:

$$\mathbf{y}_n \perp \mathbf{x}_{0:n-1} \middle| \mathbf{x}_n \qquad (2.44)$$

From the memoryless channel assumption in (2.44), (2.41) becomes:

$$p\left(\mathbf{y}_n \,|\, \mathbf{x}_{1:n}, \mathbf{y}_{1:n-1}\right) = p\left(\mathbf{y}_n \,|\, \mathbf{x}_n\right) \tag{2.45}$$

and from the Markov state assumption in (2.43), (2.42) becomes:

$$p\left(\mathbf{x}_n \,|\, \mathbf{x}_{0:n-1}, \mathbf{y}_{1:n-1}\right) = p\left(\mathbf{x}_n \,|\, \mathbf{x}_{n-1}\right) \tag{2.46}$$

Substituting (2.40), (2.45) and (2.46) into (2.37) yields:

$$w_n^{(i)} \propto w_{n-1}^{(i)} \frac{p\left(\mathbf{y}_n \,\middle|\, \mathbf{x}_n^{(i)}\right) p\left(\mathbf{x}_n^{(i)} \,\middle|\, \mathbf{x}_{n-1}^{(i)}\right)}{q\left(\mathbf{x}_n^{(i)} \,\middle|\, \mathbf{x}_{0:n-1}^{(i)}, \mathbf{y}_{1:n}\right)} \tag{2.47}$$

A final choice for the proposal distribution is to select it dependent only on the previous state and the current measurements, i.e. $q\left(\mathbf{x}_n \,|\, \mathbf{x}_{0:n-1}, \mathbf{y}_{1:n}\right) = q\left(\mathbf{x}_n \,|\, \mathbf{x}_{n-1}, \mathbf{y}_n\right)$. The weight update equation for the particle filters then becomes:

$$w_n^{(i)} \propto w_{n-1}^{(i)} \frac{p\left(\mathbf{y}_n \,\middle|\, \mathbf{x}_n^{(i)}\right) p\left(\mathbf{x}_n^{(i)} \,\middle|\, \mathbf{x}_{n-1}^{(i)}\right)}{q\left(\mathbf{x}_n^{(i)} \,\middle|\, \mathbf{x}_{n-1}^{(i)}, \mathbf{y}_n\right)} \tag{2.48}$$

This is the Sequential Importance Sampling (SIS) particle filter, aka CONDENSATION algorithm. One iteration of the SIS particle filter recursion involves the following two steps:

- Draw the updated particles $\mathbf{x}_n^{(i)}$ from the proposal distribution $q\left(\mathbf{x}_n \,\middle|\, \mathbf{x}_{n-1}^{(i)}, \mathbf{y}_n\right)$, after conditioning it upon the previous samples $\mathbf{x}_{n-1}^{(i)}$ and the current measurement $\mathbf{y}_n$.
- Obtain the current weights $w_n^{(i)}$ by updating the weights $w_{n-1}^{(i)}$ of the previous iteration. To do so the measurement model $p\left(\mathbf{y}_n \,|\, \mathbf{x}_n\right)$, the object model $p\left(\mathbf{x}_n \,|\, \mathbf{x}_{n-1}\right)$ and the proposal distribution $q\left(\mathbf{x}_n \,|\, \mathbf{x}_{n-1}, \mathbf{y}_n\right)$ are evaluated at the newly drawn samples $\mathbf{x}_n^{(i)}$, the previous ones $\mathbf{x}_{n-1}^{(i)}$ and the current measurement $\mathbf{y}_n$, as in (2.48).

Once the updated particles are obtained, the state can be estimated either using MAP or MMSE estimation. MAP estimation of (2.8) simply selects the particle with the largest weight and sets the state equal to its sample value. The MMSE estimation utilises all particles. The averaging of (2.9) is equivalent to the sum of the particle samples, weighted by the particle weights, as shown in (2.30).

The two steps of one iteration of the SIS particle filter are depicted schematically in Fig. 2.9.

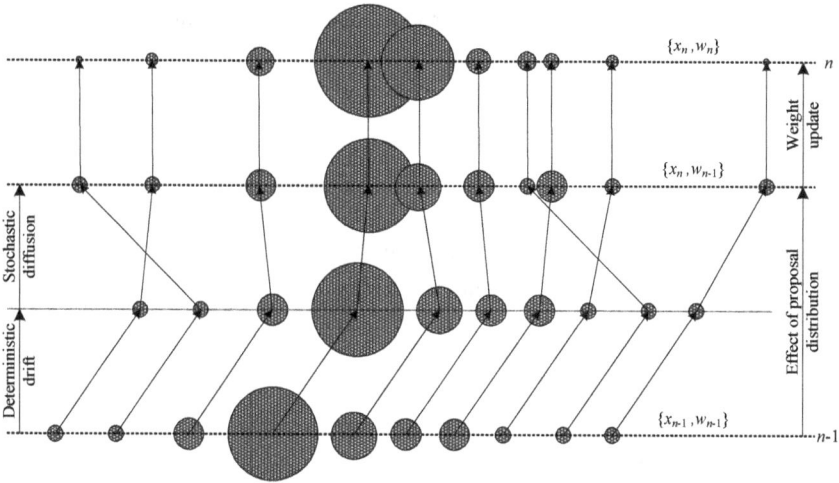

Fig. 2.9 One iteration of the SIS particle filter, involving drawing new samples from the proposal distribution conditioned upon the previous samples and the current measurement, and the update of their weights based on (2.48).

2.5.1 *Object models*

Typical radar targets tend to move in predefined ways that allow usage of strict object models. For example, there are certain sets of manoeuvres an aeroplane can perform. Humans on the other hand, especially when moving indoors, tend to do so unpredictably. Hence a suitable object model for tracking people should make only a weak assumption for state evolution. The object model should also provide for lock recovery after target loss (due to erratic motion or occlusion). These goals are met by the object model of [Pérez *et al.* (2004)]. Motion smoothness is guaranteed by a Gaussian random walk component, while target recovery is aided by a uniform component:

$$p\left(\mathbf{x}_n \left| \mathbf{x}_{n-1} \right.\right) = (1 - \beta_u)\, N\left(\mathbf{x}_n \left| \mathbf{x}_{n-1}, \mathbf{C_x} \right.\right) + \beta_u U\left(\mathbf{x}_n \left| V \right.\right) \qquad (2.49)$$

where $N\left(\mathbf{x} \left| \mu, \mathbf{C} \right.\right)$ is the multivariate Gaussian distribution with mean vector μ and covariance matrix $\mathbf{C}$, $U\left(\mathbf{x} \left| V \right.\right)$ is the uniform distribution in some volume V of the state-space, and β_u is the relative weight of the uniform distribution in the object model. β_u is usually small, allowing only a few of the particles to abandon the smooth motion of the random walk component and embark into spanning V trying to reacquire targets that might be lost.

Sampling distributions is the process of drawing random vectors of a particular PDF. To draw samples from multimodal distributions, one needs to consider the underlying distributions, as shown in example 2.5.

Example 2.5. Sampling bimodal object models.

To draw samples from the object model of (2.49), one has to consider the Gaussian distribution of the random walk component and the uniform distribution of the lock recovery component. Subsequently, one has to sample from both with a relative frequency dictated by b_u.

To sample the multivariate Gaussian distribution $N(\mathbf{x}\,|\mu, \mathbf{C})$ with mean vector μ and covariance matrix $\mathbf{C}$, one generates random vectors whose elements are uncorrelated zero-mean unity-variance Gaussian random numbers and transforms them to produce the wanted distributions by multiplying the random vectors with $\mathbf{C}^{1/2}$ and adding μ. To do this in MATLAB, assume that the dimensions of the Gaussian vectors is $\mathbf{n}$ and one draws $\mathbf{Np}$ of them, while the mean vector is $\mathbf{m}$ and the covariance matrix is $\mathbf{C}$. The following MATLAB code then generates the random vectors stacked side-by-side into the matrix $\mathbf{xg}$ of size $n \times N_p$:

```
xg=C^(1/2)*randn(n,Np)+repmat(m,1,Np);
```

To sample the uniform distribution $U(\mathbf{x}\,|V)$ in some volume V of the state-space, one generates random vectors whose elements are uncorrelated uniform random numbers in $[0,1]$. The volume V is specified by the lower and upper limits on the different dimensions, $\left\{L_{low}^{(i)}, L_{high}^{(i)}\right\}_{i=1}^{n}$. The limits are stored in the MATLAB matrix $\mathbf{L}$ which comprises a column of the $\mathbf{n}$ lower and another column of the upper limits. The following MATLAB code then generates the random vectors $\mathbf{xu}$:

```
xu=(rand(n,Np).*repmat(L(:,2)-L(:,1),1,Np)+repmat(L(:,1),1,Np));
```

The random vectors distributed as in the multimodal PDF of the object model in (2.49) are drawn from the two constituent distributions with relative frequency b_u for the uniform component and $1 - b_u$ for the Gaussian. This is easily implemented for every sample by means of a uniform random variable r in the range $[0,1]$, by comparing its value to b_u and then drawing the random vector from the uniform PDF if $r < b_u$, or else by drawing it from the Gaussian PDF. In MATLAB this is implemented as:

```
bu=.25;
idx=find(rand(1,Np)<bu);
x=xg;
x(:,idx)=xu(:,idx);
```

An example is given in Fig. 2.10, for a one-dimensional state, with obviously `n=1`, and also `Np=1e5`, `L=[-6 6]`, `C=0.5`, `m=1` and `bu=0.25`.

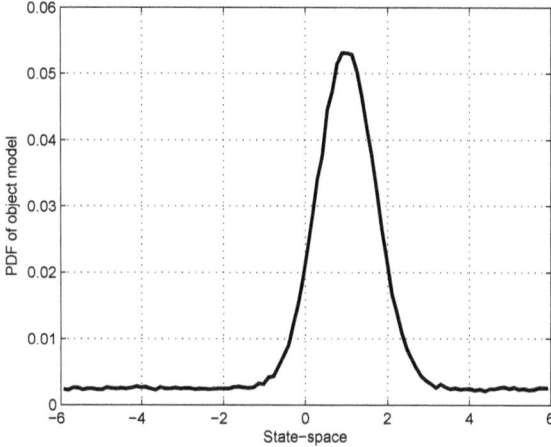

Fig. 2.10   Bimodal PDF of the form of the object model in (2.49), estimated from the samples drawn as discussed in example 2.5. The uniform and the Gaussian components are clearly visible.

∎

2.5.2   *Object model for proposal distribution*

A common choice for the proposal distribution is the object model itself , $q\left(\mathbf{x}_n \mid \mathbf{x}_{n-1}, \mathbf{y}_n\right) = p\left(\mathbf{x}_n \mid \mathbf{x}_{n-1}\right)$. In this case the weight update in (2.48) simplifies to:

$$w_n^{(i)} \propto w_{n-1}^{(i)} p\left(\mathbf{y}_n \mid \mathbf{x}_n^{(i)}\right) \qquad (2.50)$$

This choice simplifies the particle filter, as the particles are obtained by first drawing samples from the object model PDF and then updating the weights by evaluating the measurement model PDF at the drawn sample values.

The implementation of an SIS particle filter employing the object model for proposal distribution, for the one-dimensional case, is detailed in example 2.6.

Example 2.6. SIS particle filter for 1D motion.

A target to be tracked is moving along a line with known speed. Hence the hidden states to be estimated are given by $s_n = s_{n-1} + \Delta x$, where $\Delta x = 0.5$ is known and the initial state $s_0 = 4$ is unknown. The states are observed by some measurements y_n which are Gaussian random variables with mean s_n and variance $\sigma_y^2 = 1$. The goal is to estimate the hidden states by means of an SIS particle filter utilising the object model as proposal distribution.

The hidden states and the noisy measurements are generated by the following MATLAB code:

```
so=4;              % Initial hidden state
Dx=.5;             % Deterministic drift, x(n)=x(n-1)+Dx;
sy2=1;             % Variance of Gaussian measurement model
N=20;              % Time steps to simulate
s=so+Dx*(0:N-1);
y=s+sqrt(sy2)*randn(1,N);
```

The particle filter tracker employs $N_p = 30$ particles $\left\{ x_n^{(i)}, w_n^{(i)} \right\}_{i=1}^{N_p}$, represented in MATLAB by x and w. At initialisation the $x_0^{(i)}$ are uniformly distributed around y_0 at a length of 3 units, with weights $w_0^{(i)} = 1/N_p$. The state estimates x_n, denoted in MATLAB by xn, are obtained as the mean of the target distribution, employing (2.30):

```
Np=30;                 % Number of particles
x=3*rand(1,Np)+y(1)-1.5;
w=ones(1,Np)/Np;
xn=[x*w' zeros(1,N-1)];
```

Note that the only reason the history of state estimations for the N simulated time steps is kept, is for performance evaluation.

Following the initialisation, the N time steps are simulated in a for-loop, by drawing new particles $x_n^{(i)}$ and updating the previous weights $w_{n-1}^{(i)}$.

The object model to draw each new particle $x_n^{(i)}$ from is similar to the one in example 2.5:

$$p\left(x_n \left| x_{n-1}^{(i)} \right. \right) = (1 - \beta_u) N\left(x_n \left| x_{n-1}^{(i)} + \Delta x, \sigma_x^2 \right. \right) + \beta_u U\left(x_n \left| V \right. \right) \quad (2.51)$$

The Gaussian component $N\left(x_n \left| x_{n-1}^{(i)} + \Delta x, \sigma_x^2 \right. \right)$ of the object model to draw each particle i has a mean value obtained as the particle values at the previous instant, $x_{n-1}^{(i)}$, shifted by Δx. Its variance is $\sigma_x^2 = 1$. The uniform component $U\left(x_n \left| V \right. \right)$ of the object model spans a volume V (actually a line in the 1D case) in the range $[0, x_{max}]$, with $x_{max} = 25$ and has relative weight $\beta_u = 0.1$. Drawing the particles from this object model is similar to example 2.5.

The measurement likelihood function is a Gaussian:

$$p\left(y_n \,|x_n\,\right) = N\left(y_n \,|x_n, \sigma_y^2\right) \tag{2.52}$$

and is evaluated at the particles to perform weight update as in (2.50):

```
py=zeros(1,Np);
for i=1:Np
    py(i)=normpdf(y(n),x(i),sqrt(sy2));
end
w=w.*py;
w=w/sum(w);
```

The final step in the loop is to obtain the new state estimate. The MMSE estimation is implemented as:

```
xn(n)=x*w';
```

According to the MAP estimation, the state is:

```
[m i]=max(w);
xn(n)=x(i);
```

A run of the particle filter for the few first time steps is given in Fig. 2.11. Most of the particle trajectories are concentrated around the measurements and the hidden states. For most time steps, the particle filter provides state estimations that are closer to the hidden ones than the measurements. There are some trajectories though, albeit with very small weights, that at some point in time are not updated by the usual smooth Gaussian random walk component but by the sparsely exercised uniform lock recovery component that scans the whole of the state-space. This forces the trajectories away from the hidden states. Occasionally these trajectories are supported by measurements, like the one in time step 7. This is rare; usually these trajectories are not supported by some measurement, so their weights diminish to negligible values.

Particle filters do not offer the advantage of the analytical solution of the Kalman filters, which is constant given the measurements and the models. The filtering process in particle filters yields numerical approximations, varying across different runs, as random numbers are drawn from proposal distributions. Hence the particle filter is run 5,000 times for 40 time steps. The median RMS state estimation error is 0.82, while the RMS error of the measurements equals their standard deviation ($\sqrt{\sigma_y^2}$) and hence is unity.

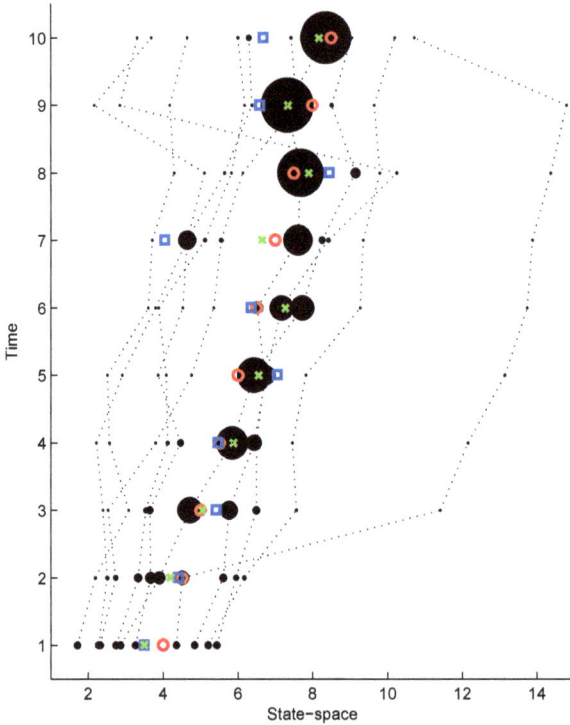

Fig. 2.11   Particles of the SIS particle filter in example 2.6 for the first few time steps. The hidden states to be estimated are shown in red circles, the observable measurements in blue squares and the estimated states in green "x". The filter estimates the hidden states better than the measurements.

Finally note that the state estimation provided by the particle filter improves by increasing the number of particles N_p. In the particular example using only 20 particles results in state estimation that is consistently worse than just trusting the measurements, while increasing their number to the computationally very demanding $N_p = 150$, drops the estimation error to 0.70.

◼

2.5.3   *Degeneracy of SIS particle filter*

One final observation is due from example 2.6 and Fig. 2.11: as time goes by, the weights concentrate on increasingly fewer particles. After a sufficient number of time steps, all but one particle have negligible weights. This leads to spending resources updating particles that have no effect in the filtering process, and to reducing the effective number of particles. This is the sample degeneracy problem. Degeneracy of the SIS filter cannot be addressed by using an excessive number of particles; its onset will just be delayed, since it has been shown in [Doucet *et al.* (2000)] that the variance of the weights increases over time. Instead it is addressed in two ways:

- Carefully choosing the proposal distribution, as discussed in section 2.5.4.
- Employing resampling, as discussed in section 2.5.5.

2.5.4 *Optimum proposal distributions*

The optimum proposal distribution $q\left(\mathbf{x}_n \left| \mathbf{x}_{n-1}, \mathbf{y}_n \right.\right)$ is known [Doucet *et al.* (2000)] to be the posterior:

$$q_{opt}\left(\mathbf{x}_n \left| \mathbf{x}_{n-1}, \mathbf{y}_n \right.\right) = p\left(\mathbf{x}_n \left| \mathbf{x}_{n-1}, \mathbf{y}_n \right.\right) = \frac{p\left(\mathbf{y}_n \left| \mathbf{x}_n \right.\right) p\left(\mathbf{x}_n \left| \mathbf{x}_{n-1} \right.\right)}{p\left(\mathbf{y}_n \left| \mathbf{x}_{n-1} \right.\right)} \quad (2.53)$$

Substituting (2.53) into the weight update eq. (2.48), the optimum weights become:

$$w_n^{(i)} \propto w_{n-1}^{(i)} p\left(\mathbf{y}_n \left| \mathbf{x}_{n-1}^{(i)} \right.\right) = w_{n-1}^{(i)} \int p\left(\mathbf{y}_n \left| \mathbf{x}_n \right.\right) p\left(\mathbf{x}_n \left| \mathbf{x}_{n-1}^{(i)} \right.\right) d\mathbf{x}_n \quad (2.54)$$

Note that given the previous particles $\left\{\mathbf{x}_{n-1}^{(i)}, w_{n-1}^{(i)}\right\}_{i=1}^{N_p}$ at time $n-1$, the new weights do not depend on the new samples drawn from the optimum proposal distribution, hence the variance of the updated weights conditioned upon the previous particles is zero, i.e. it is minimised.

Unfortunately, a particle filter with such a proposal distribution would require sampling from it and evaluating the integral in the weight update of (2.54), neither of which is possible in the general case. Both tasks are possible if the state space is discrete or the proposal distribution is Gaussian. Hence a suboptimal choice is sought. The probability of the current measurement given the previous state, $p\left(\mathbf{y}_n \left| \mathbf{x}_{n-1} \right.\right)$, in the denominator of the optimum proposal distribution (2.54) is just a normalising constant, yielding:

$$q_{opt}\left(\mathbf{x}_n \left| \mathbf{x}_{n-1}, \mathbf{y}_n \right.\right) \propto p\left(\mathbf{x}_n \left| \mathbf{x}_{n-1} \right.\right) p\left(\mathbf{y}_n \left| \mathbf{x}_n \right.\right) \quad (2.55)$$

The most widely used approximation to the optimum proposal distribution is to use the object model, i.e. only the first term of (2.55), as already discussed in section 2.5.2, leading to the SIS particle filter. In order to avoid the need for large number of particles N_p, the proposal distribution should match the posterior. To do so, the second term of (2.55), i.e. the measurement model, should also be taken into account. Options for measurement-assisted proposal distributions are presented in section 4.11.3.

2.5.5 *Resampling*

Resampling [Doucet *et al.* (2000)] is a stochastic process during which particles with small weights are likely to be replaced by replicas of particles with large weights. In the long run, particles with small weights are eliminated, allowing the update process to concentrate on particles with large weights. To do so, any resampled sample $\mathbf{x}_r^{(i)}$ has the value of the j-th sample in the original particles with probability $w^{(j)}$, $Pr\left(\mathbf{x}_r^{(i)} = \mathbf{x}^{(j)}\right) = w^{(j)}$. After resampling the weights are reset to $1/N_p$.

There are many different resampling algorithms in the relevant literature [Douc *et al.* (2005)]. Systematic resampling [Kitagawa (1996)] is selected amongst them due to its performance, simplicity and speed. The cumulative sum of the weights $c^{(j)} = \sum_{k=1}^{j} w^{(k)}$ is scanned from a starting point u_1 randomly selected in the interval $[0, 1/N_p]$ and incremented by $1/N_p$ steps. The process is started from the first resampled sample value $\mathbf{x}_r^{(1)}$ and the first original sample value $\mathbf{x}^{(1)}$. For the i-th resampled sample value, increase j until the j-th cumulative weight c_j is larger than or equal $u_1 + (i-1)/N_p$. Then set $\mathbf{x}_r^{(i)} = \mathbf{x}^{(j)}$.

Systematic resampling is depicted in Fig. 2.12. If for some i, j is incremented more than once (as for the second original particle in Fig. 2.12), or if for $i = N_p$ it is $u_1 + (N_p - 1)/N_p \le c_j$ (as for the tenth original particle in Fig. 2.12), then the respective sample values $\mathbf{x}^{(j)}$ do not appear in the resampled set. On the other hand, if i is increased without increasing j, then the respective sample value $\mathbf{x}^{(j)}$ appears multiple times in the resampled set (as for the seventh original particle in Fig. 2.12).

Systematic resampling is applied in example 2.7 for improving the standard deviation estimation of the approximation of a PDF with particles.

Example 2.7. Improve standard deviation estimation of particles by resampling.

Assume Np one-dimensional particles drawn uniformly in the range $[-a, a]$, representing a Gaussian distribution of mean m and standard deviation σ, where $m + 3\sigma < a$ and $m - 3\sigma > -a$. The weights of the particles

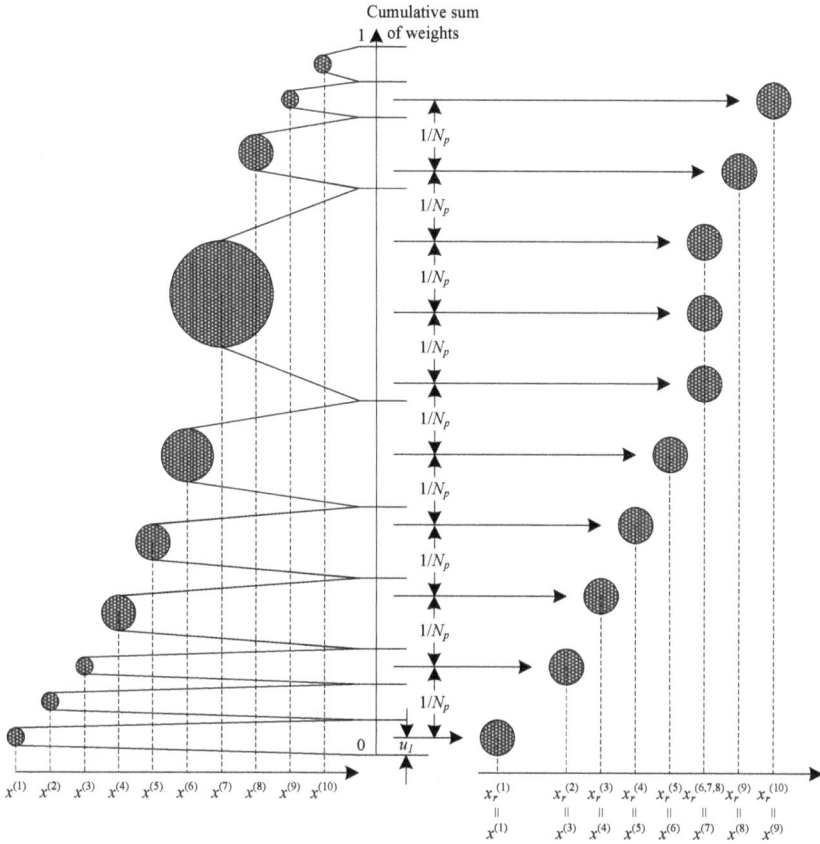

Fig. 2.12 Systematic resampling of the original sample values $\mathbf{x}^{(j)}$ into $\mathbf{x}_r^{(i)}$, shown at the horizontal axes. The vertical axis depicts the cumulative sum of the weights, which is resampled at regular intervals $1/N_p$, starting at the random choice u_1.

are given by evaluating the Gaussian distribution at the sample values $x^{(i)}$:

```
a=5;
x=(rand(1,Np)-.5)*2a;
m=1;s=1;
w=normpdf(x,m,s);w=w/sum(w);
```

Although the resulting particles offer a good estimation of the mean value, they do not do so for the standard deviation. This estimate is obtained as the square root of the variance estimate:

$$var[x] = E\left[(x - E[x])^2\right] = E\left[x^2\right] - E[x]^2 \qquad (2.56)$$

Using (2.29) and (2.28), (2.56) yields:

$$var[x] \simeq \sum_{i=1}^{N_p} w^{(i)} \left(x^{(i)} \right)^2 - \sum_{i=1}^{N_p} w^{(i)} x^{(i)} \qquad (2.57)$$

The RMS estimation error of the standard deviation as a function of the number of particles is shown in Fig. 2.13. This estimation error decreases as the number of particles increases, up to some saturation point.

In order to evaluate if resampling results in a better estimation, systematic resampling is implemented as follows:

```
c=cumsum(w);
u=rand(1)/Np;
xr=zeros(size(x));
j=1;
for i=1:Np
    while c(j)<u
        j=j+1;
    end
    xr(:,i)=x(:,j);
    u=u+1/Np;
end
x=xr;
w=repmat(1/Np,1,Np);
```

The RMS error of the standard deviation estimation with resampling is also shown in Fig. 2.13. The error with resampling is smaller than that without and it consistently declines as the number of particles increase, without any sign of saturation as in the case without resampling.

■

2.5.6 *Sequential Importance Resampling particle filter*

No particle filter can be used in practice without some form of resampling. The Sequential Importance Resampling (SIR) particle filter[Arulampalam *et al.* (2002)], aka Sampling Importance Resampling or Bootstrap particle filter, uses the object model for proposal distribution such as the SIS particle filter of example 2.6, but also involves resampling at each iteration. Note that this choice for a proposal distribution is not optimal, since it only includes the first term of the posterior in (2.53), ignoring the influence of the second term, the measurement model. The three steps of the SIR particle filter are depicted schematically in Fig. 2.14.

When particles are resampled, some of them then have the same sample values. This is the sample impoverishment problem. Usually it is alleviated

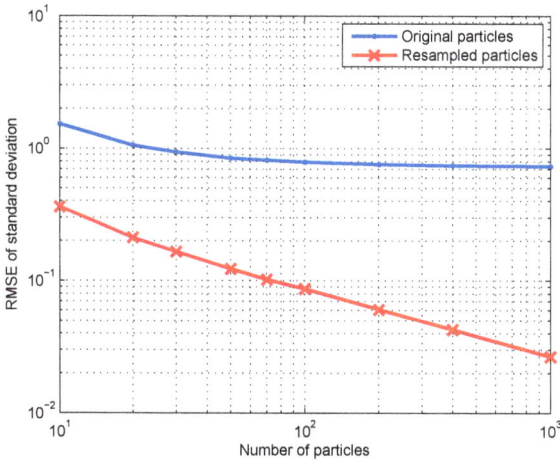

Fig. 2.13 RMSE of the standard deviation estimation in example 2.7, with and without resampling, as a function of the number of particles used.

by the process noise of the object model that scatters these particles at the next time step. Sample impoverishment can be quite severe in cases where the process noise is very small.

A note is needed about state estimation when resampling is used. The estimate either based on MAP or MMSE is obtained prior to resampling. The SIR particle filter is compared to the SIS in example 2.8.

Example 2.8. Comparison of the SIR and SIS particle filters.

Returning to the SIS particle filter of example 2.6 and turning it into an SIR one by adding resampling at each iteration, the state estimation improves drastically, as shown by the solid lines in Fig. 2.15. The RMS state estimation error is measured as in example 2.6. For comparison, the estimated performance of the filter without resampling (i.e. of the SIS particle filter of example 2.6) is also included in dashed lines. When resampling is used in this simple 1D example, there is no need for more than 30 particles.

A run of the SIR PF for the few first time steps is given in Fig. 2.16. Compared with the same run (same hidden states, measurements and models) without resampling shown in Fig. 2.11, the resampled trajectories are more densely located around the hidden states and none of them receives negligible weight. This is because any trajectory that strays away, unless soon supported by a measurement, receives small weight and is likely to

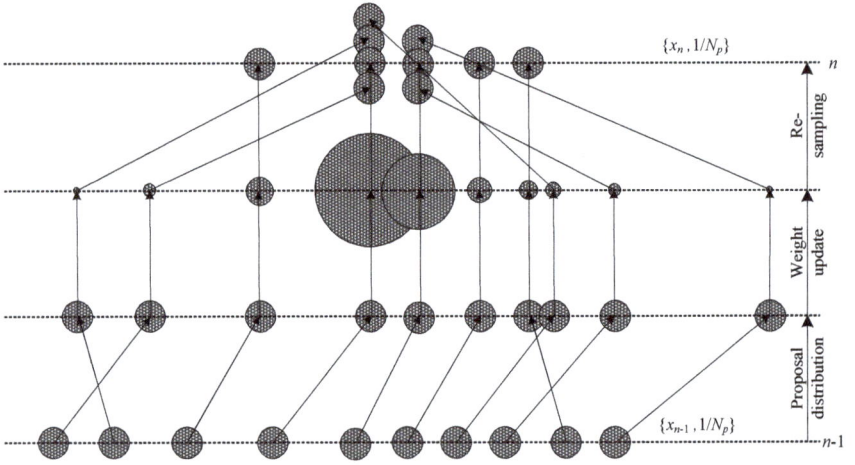

Fig. 2.14 Operation of the SIR particle filter, including drawing new samples from the object model, updating the weights according to (2.50) and resampling to avoid the degeneracy problem.

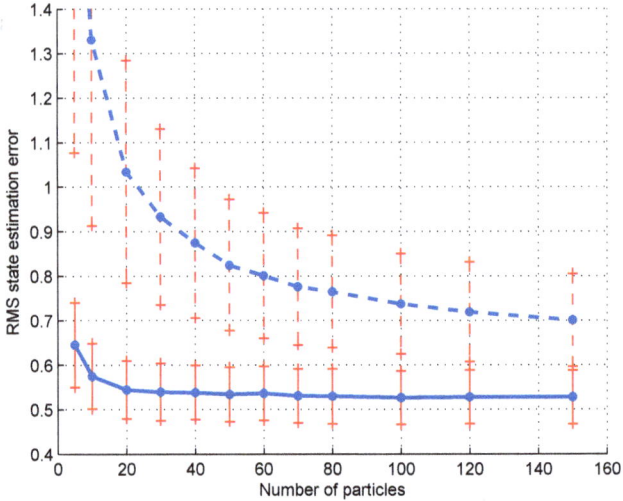

Fig. 2.15 Median and inter-quartile range of the RMS state estimation error of the particle filter of example 2.6 for 40 time steps, without (dashed lines) and with resampling (solid lines). The latter is the SIR particle filter. The inter-quartile range is represented by the length of the vertical bars centred on the median.

be resampled to the sample value of some highly weighted particle (see for
example the leftmost particle in the transition from the second to the third
time step).

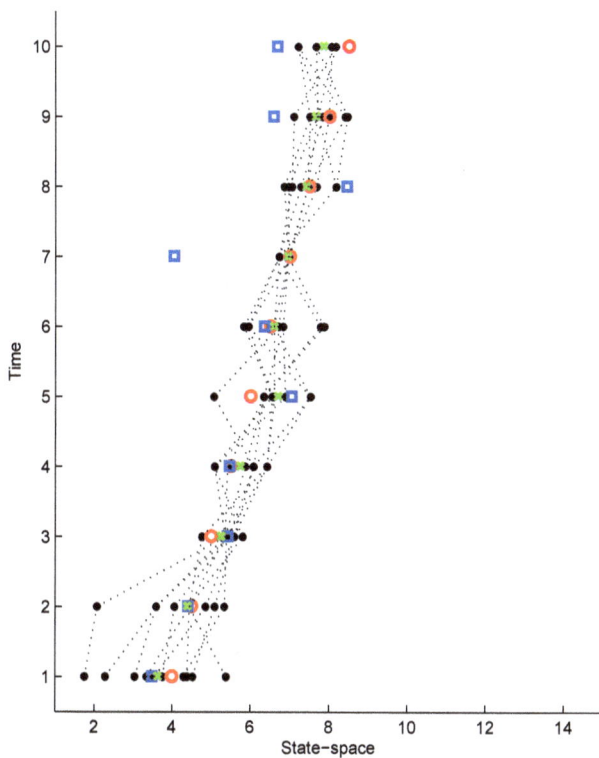

Fig. 2.16 Particles of the SIR particle filter in example 2.6 for the first few time steps.
The hidden states to be estimated are shown in red circles, the observable measurements
in blue squares and the estimated states in green "x". To be compared to the trajectories
without resampling in Fig. 2.16.

2.6 Conclusions

Particle filters have been introduced as a numerical approximation to the
recursive Bayesian filtering problem, without the linearity and Gaussianity
constraints necessary for the exact solution of the Kalman filters. The

problem of degeneracy in particle filters has been identified, leading to the SIR particle filter. Simple 1D tracking examples have been used throughout the chapter. In the following chapters complete tracking systems are built, based on the SIR or more elaborate particle filters.

Chapter 3

Audio Tracking

3.1 Introduction

The problem of locating and tracking an acoustic source using microphone arrays is subject to a series of problems imposed by the acoustic environment. These include multi-path effects, ambient noise, multiple and competing speakers and pauses in human speech. A solution to the problem is crucial for a series of human computing applications. Typical examples are new video-conferencing systems where cameras are steered towards the person that is speaking [Brandstein and Ward (2001); Nishiura *et al.* (2001)] and also autonomous recording systems [Azodolmolky *et al.* (2005)] where only the best view camera stream is chosen to be recorded. Further applications where person localisation and tracking become important are discussed in Chapter 6.

The combination of algorithms for solving both the localisation and tracking problems is called an Acoustic Source Localisation and Tracking (ASLT) system. The acoustic source we are attempting to localise and track in a room generates a sound field which varies in character at different places in the enclosure. In order to understand the limiting factors of ASLT it is thus important to identify these different descriptors.

Assuming a single source and microphone in the enclosure an audio recording would include the signal as it arrives from the shortest path i.e. the direct line connecting the source and the microphone (known as the direct-path). Shortly after, the microphone would record delayed and attenuated versions of the same signal, the reflections, as these bounce off walls and objects in the enclosure (travelling through paths known as the reverberant-paths). Additionally, the recordings are subject to interfering noise sources that degrade the quality of the signals further.

45

A microphone at a location near the sound source is said to be in the near-field of the source. In the near-field, the signal recorded is in practice determined by the source itself i.e. the signal travelling through the direct-path. A little further from the source, the sound field begins to be affected by the medium and the surroundings. The part of the sound field outside the near-field is called the far-field. In general the ASLT problem becomes easier when considering near-field conditions since the effect of reverberation and noise is less significant. Nevertheless, in most cases microphones are at the far-field of the sound source and receive signals travelling from both direct and reverberant paths as well as signals from noise sources.

Time Delay Estimation (TDE) methods, such as the Generalized Cross Correlation (GCC) [Knapp and Carter (1976)], are considered the most popular building component for ASLT. TDE methods estimate the direction from which sound is arriving from, based on the time difference between pairs of microphone recordings. The ASLT system then combines such estimates to return the actual location. The practical, and in many ways severely restricting, disadvantage of this approach is that if the system is used in reverberant environments the location estimate could be spurious due to ensuing reflections. Methods to limit such effects by providing more robust TDE can be found in [Talantzis *et al.* (2005); Brandstein *et al.* (1997); Benesty *et al.* (2007)]. In the following paragraphs we will discuss one such method and analyse the benefits offered when compared with GCC.

A second class of ASLT algorithms relies on beamforming (BF)[Bolic *et al.* (2002)]. In this case, microphones are organised to focus at specific locations in the room and measure the frequency-averaged power. The assumption is that the BF output power would be large at the source location. Again, the problems imposed by the environment result in the generation of erroneous maxima [Peled and Rafaely (2008)]. In this Chapter we will discuss a simple power-based BF as well as a variant that operates better in reverberant and noisy conditions.

Original TDE and BF methods relied on statistics provided during the current frame only. In this way information about the motion patterns of the acoustic source were neglected. Particle filters (PF) exploit this property better by assuming that the localisation measurement (TDE or BF) will follow some dynamical model from frame-to-frame while the erroneous estimates will have no consistency in time [Cour (2007); Arulampalam *et al.* (2002)]. Such consistent behaviour is present in human walking where people move between points with a certain speed and acceleration that can be modelled to a good extent. An analytical investigation of the performance of PFs when used in conjunction with GCC and simple forms of BF can be found in [Ward *et al.* (2003)]. In the remainder of the chapter we will perform a similar analysis for the case where more robust algorithms for TDE and BF are applied in a modified PF framework.

In the context of ASLT it is also important to note that human speech utterances are non-stationary [Lehmann and Johansson (2007)]. Thus, the acoustic signals have potentially silent gaps during which speakers might also move. During those pauses the ASLT system would keep updating the source location as if the speaker were active. The algorithms would therefore lose track of the true source and possibly "get trapped" in an erroneous location generated by background noise or reverberation. The employment of a voice activity detector (VAD) can potentially generate an improvement in performance since source location estimates will only be updated during periods that the acoustic source is active. Nevertheless, the addition of a VAD module in an ASLT system will generate additional computational load and thus limit its potential in a real-time implementation. Additionally, in a realistic scenario the ASLT system would have to cope with interchanging speakers that may reside far apart. To this end, this chapter also discusses a VAD algorithm to support the ASLT system.

Keeping in mind all of the above factors a block diagram of an ASLT system can be seen in Fig. 3.1. There are scenarios were the VAD and detection modules can be changed in order i.e. first detect voice activity and then activate the detector mechanism.

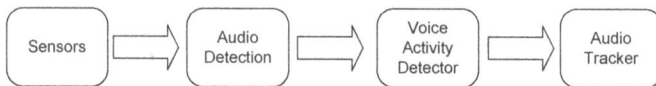

Fig. 3.1   Block diagram of an ASLT system.

3.2   Multiple audio sensors

Multi-path effects and ambient noise affect drastically the performance of any Audio Signal Processing system. In the case of ASLT these effects result in inaccurate source location estimates since reflections are mistaken for the actual source while noise possibly suppresses spatial information further. It is today generally accepted in current literature that removal of these effects can be greatly enabled by using multiple microphones organised in microphone arrays. The signals from these microphones, combined in a certain way, increase the directivity of the array and potentially reduce the captured noise and reverberation.

The purpose of the following paragraphs is to introduce the system-model used when employing microphone arrays. In addition we will discuss the tools used and the assumptions made in order to simulate acoustic reverberant environments in order to optimise the design of real-world systems.

3.2.1 *System model*

An audio source localisation and tracking (ASLT) system considers M microphones in a multi-path environment. Assuming a single source, the discrete signal recorded at the mth microphone ($m = 1, 2, \ldots, M$) at time k is:

$$x_m(k) = h_m(k) * s(k) + n_m(k), \qquad (3.1)$$

where $s(k)$ is the source signal, $h_m(k)$ is the room impulse response between the source and m^{th} microphone, $n_m(k)$ is additive noise and $*$ denotes convolution. The length of $h_m(k)$, and thus the number of reflections, is a function of the reverberation time T_{60} (defined as the time in seconds for the reverberation level to decay to 60 dB below the initial level) of the room and demonstrates one of the main problems when attempting to track an acoustic source. As mentioned earlier, this is because when the system is used in reverberant environments the source location estimate could occur in a spurious location created by the ensuing reflections.

Given that ASLT systems typically operate in real-time, we assume that data at each sensor m are collected over t frames of data $\mathbf{x}_m^{[t]} = [x_m(tL), x_m(tL + 1), \ldots, x_m(tL + L - 1)]$ of L samples. So, at frame t the representation of the microphone data is formed as follows:

$$\mathbf{y}_{1:t} = \begin{bmatrix} \mathbf{x}_1^{[1]} & \mathbf{x}_1^{[2]} & \cdots & \mathbf{x}_1^{[t]} \\ \mathbf{x}_2^{[1]} & \mathbf{x}_2^{[2]} & \cdots & \mathbf{x}_2^{[t]} \\ & & \vdots & \\ \mathbf{x}_m^{[1]} & \mathbf{x}_m^{[2]} & \cdots & \mathbf{x}_m^{[t]} \end{bmatrix} \qquad (3.2)$$

Thus, the role of an ASLT system is to use the representation of eq. (3.2) in order to identify the Cartesian coordinates of the active acoustic source.

Older localisation systems ignored the concatenation of frames as seen in eq. (3.2) and attempted to estimate the source location using data from the current frame only i.e. using a single column from eq. (3.2).

Example 3.1. The following MATLAB code-fragment frames sampled data contained in a vector x of length L_x using an increment inc and a window win of length L_win. The frame size is L samples.

```
no_frames = fix((L_x-L+inc)/inc);
x_f = zeros(no_frames,L);
f = inc*(0:(no_frames-1)).';
s = (1:L);
x_f = x(f(:,ones(1,L))+s(ones(no_frames,1),:));
if (L_win > 1)
    x_f = x_f .* win(ones(no_frames,1),:);
end
```

■

For the remainder of this Chapter we will concentrate on the frequency domain representation of (3.2). Thus, we take a set of frames of L samples from all m microphones $\mathbf{x}_m^{[t]}$ and convert them into the frequency domain using an L-point Short Time Fourier Transform (STFT) to get $\mathbf{X}_m^{[t]} = [X_m(\omega_0), X_m(\omega_1), \ldots, X_m(\omega_{L-1})], m = 1, 2, \ldots M$. In this case, ω_l denotes the lth discrete frequency bin with $l = 0, 1, \ldots L - 1$.

Example 3.2. Converting a single frame of data x_m from microphone m into its L-point frequency-domain version X_m is a matter of a single command in MATLAB:

```
X_m = fft(x_m,L);
```

■

3.2.2   *The image model*

When working towards the solution of Acoustical Signal Processing problems it often becomes necessary to find ways of modelling the effect of the acoustic enclosure upon the source signal i.e. quantify an impulse response (IR) between an acoustic source and a microphone. This facilitates the process of using simulations as a design tool for the subsequent employment of the algorithms in real-world scenarios.

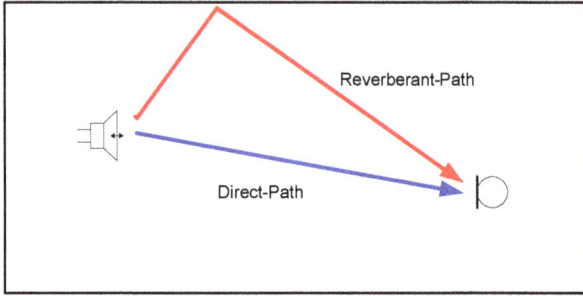

(a) Simple model of reverberation in a room.

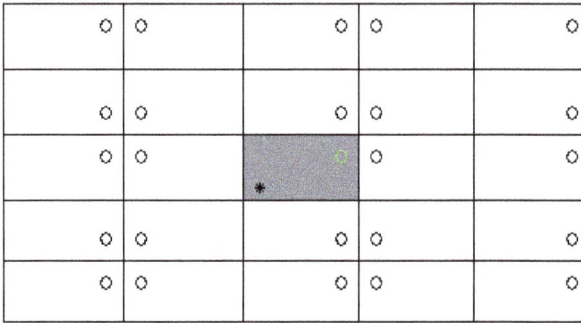

(b) Higher order reflections represented by image sources.

Fig. 3.2   Schematic representation of the image model.

Mathematically, the sound propagation is described by the wave equation. Thus, an IR from a source to a microphone could be obtained by solving the wave equation. Nevertheless, the mathematics behind the solution make this approach impractical especially in the case where the environment changes dynamically (e.g. the source moves) and/or one would like to simulate a large number of impulse responses. The mathematics of the solution are out of the scope of this book and can be found analytically in [Nelson and Elliott (1992)].

The image model [Allen and Berkley (1979)] is a convenient simulation tool for rectangular enclosures with possibly non-uniform distributions of wall absorption. It overcomes the limitations of classical statistical room acoustics [Radlović *et al.* (2000)], which assume diffuse-field conditions that are not likely at low frequencies and for unevenly distributed wall absorption. The image model can be used in order to generate a synthetic room

IR, i.e. a transfer function between a sound source and an acoustic sensor, in a given artificial environment. Once such an IR is available, reverberant audio data can be obtained by convolving the IR with a given source signal. This provides a realistic version of the sound signal as it would be recorded at a sensor in a real environment with the same properties

For the purposes of the image model, the concept of a sound ray replaces that of a wave. Rays have infinitesimal cross-section and reach a finite receiver volume. When the ray strikes a room boundary, the energy content is reduced by an amount dictated by the absorption coefficient of the specific surface. If specular reflection is assumed, then the ray leaving the surface can be assumed to emanate from an image source as far behind the plane of the room surface as the real source is in front.

The concept behind the way the image model contributes to the direct path and the reflections is conceptually simple. Assume a sound source of power Q placed near a reflecting surface. An image source will result in a power $(1 - r)Q$, where r is the absorption coefficient of the surface. The propagation path of the reflected ray is equivalent to the line from the image source to the receiving point. With more than one surface, second order image sources are generated, corresponding to rays that have suffered multiple reflections. Continuing in this way, more and more image sources are generated, while the propagation path increases and the corresponding energy contribution decreases. The number of reflections increases until the rays either arrive at a perfectly absorbent surface or their energy content becomes vanishingly small.

Schematically the described process is shown in Fig. 3.2. In the simple scenario of Fig. 3.2(a), the direct-path (represented by the shortest distance between the source and the microphone) will contribute to the IR by a single delay and an attenuation factor that is inversely proportional to the distance travelled. Similarly, a reflection contributes to the IR by a longer delay and a greater attenuation due to the fact that the path travelled by the ray is longer. Extending this to multiple images results in the situation of Fig. 3.2(b) where the ray paths are now straight lines between the image sources (denoted by circles) and the microphone (denoted by the star).

Thus, if we assume all of the walls have the same reflection properties the resulting IR between the emitting acoustic source and microphone m in an acoustic environment will be modelled by the image model as:

$$h_m(t) = \sum_{\mu=0}^{I} \frac{\beta^{\mathbf{r}_\mu}}{R_\mu} s(t - \frac{R_\mu}{c})$$ (3.3)

where I denotes the maximum number of image sources included in the model, β is the wall reflection coefficient and $\mathbf{r}_\mu$ denotes the image order i.e. the number of reflections considered for the μth image source. Finally, R_μ denotes the Euclidean distance between the source and the μth image source ($\mu = 0$ refers to the direct-path). Given the definition of T_{60} discussed in section 3.2.1 we make $h_m(t)$ long enough to sample the reverberation time i.e. $I = round(T_{60}f_s)$, where $round(.)$ denotes rounding to the closest integer.

Example 3.3. The following code-fragment [Gaubitch (2003)] is a possible implementation for simulating the impulse response of a specified room using the image model. This implementation is based on the work of [Allen and Berkley (1979)] modified according to [Peterson (1986)] in order to make it more suitable for multi-channel simulations. In this, R is an Mx3 array specifying the (x,y,z) coordinates of the M microphone(s) (metres), s is a 1x3 vector specifying the (x,y,z) coordinates of the source (metres), L is a 1x3 vector specyfying the room dimensions (x,y,z) (metres), T60 is the reverberation time (seconds), c is the speed of sound (metres/second), fs is the sampling frequency (Hz) and npts is the number of sampling points to be included in the impulse response. Finally, ht holds the resulting impulse response(s).

```
V = prod(L); S = 2*(L(1)*L(3)+L(2)*L(3)+L(1)*L(2));
alfa = 0.161*V/(S*T60); beta = sqrt(1-alfa);
N = fix(fs*0.008); Nh = fix(N/2);
win = hann(N)';
l = c/fs;
s = s/l; L = L/l; R = R/l;
M = size(R,1); ht = zeros(npts,M);

for mm = 1:M
    r = R(mm,:);
    httemp = zeros(npts+N,1);
    if (norm(s-r) < 5) ht(1,mm) = 1; continue; end
    n = ceil(npts./(2*L));
    for nx = -n(1):n(1)
    for ny = -n(2):n(2)
    for nz = -n(3):n(3)
        nr = [nx,ny,nz];
        Rr = 2*(L.*nr);
```

```
        for ll = 0:1
        for jj = 0:1
        for kk = 0:1
            np = [ll,jj,kk];
            Rp = (s-r) + (2*np.*r);
            delp = norm(Rr-Rp);
            id = fix(delp)+1;
            if (id <= npts)
                zz = sum(abs(nr) + abs(nr-np));
                gid = beta^zz/(4*pi*delp*l);
                ngid = sincfilt(delp,gid,N,Nh,win);
                httemp(id:id+N-1) = httemp(id:id+N-1) + ngid;
            end
        end
        end
        end
    end
    end
    end
    ht(:,mm) = httemp(Nh:npts+Nh-1);
end

function ngid = sincfilt(delp,gid,N,Nh,win)
d = (delp - fix(delp));
D = Nh-1 + d;
n = (0:N-1) - D;
f = sinc(n).*win;
ngid = gid*f(:);
```

∎

3.2.3 *Microphone arrays and their geometries*

A microphone array is an arrangement of multiple spatially separated microphones working together to receive a better version of a propagating wavefield. Typically, using multiple microphones becomes significant in the following scenarios:

- Locating the position of an acoustic source.
- Receiving an audio message from a distant source.
- Identify the IR of a medium through which the wavefield is propagating.

The most commonly used array geometries are uniform linear array (ULA) and uniform circular array (UCA). Nevertheless, different geometrical designs have been proposed in relevant literature, characteristic examples of which can be found in Fig. 3.3 which considers array topologies of four microphones located at Cartesian coordinates $\mathbf{m}_1, \mathbf{m}_2, \mathbf{m}_3, \mathbf{m}_4$. The four topologies depicted are denoted as T_1, T_2, T_3 and T_4. Note that topology T_4 is the only one that is not planar. It is evident that microphone arrays can be employed using more sensors and different topologies. Nevertheless, hardware and space limitations generally limit the design of arrays. According to their geometry, arrays can generally be categorised as follows:

- Linear (1D)
- Planar (2D)
- Volumetric (3D)

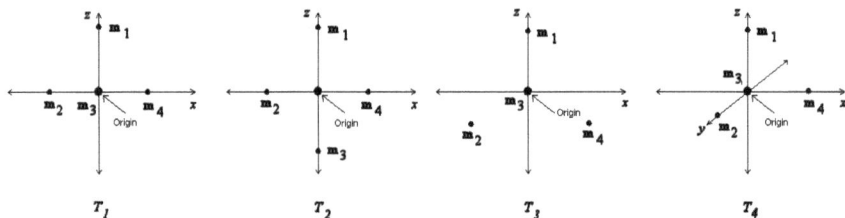

Fig. 3.3   Array topologies schematically.

The reason for grouping microphones into arrays resides in the fact that a wavefront propagating across the array of sensors is picked up by all sensors. Thus, we have multiple signals constituting the microphone array signal. In the simplest case, all components of the array signals are simply delayed replicas of the emitted signal. In the worst case, individual array signals are strongly corrupted with noise and reflections, leaving very little resemblance among them. In the context of ASLT, array processing allows us to combine all sensor outputs in some algorithmic manner so that only the original signal emitted by the source is received and its location is estimated more accurately. This is achieved by exploiting the similarity between different versions of the same signal as captured by different microphones at different spatial locations.

As noted earlier, the distance between the acoustic source to be located and the microphone arrays identifies whether our ASLT system operates in near-field or far-field conditions. In the case of near-field the source-to-array distance is comparable to array dimensions while the acoustic wavefront curvature is not negligible. Far-field conditions assume source-to-array distance that is bigger than array dimensions and planar wavefronts. Generally, in real-rooms ASLT systems can be assumed to operate in far-field conditions since microphone arrays are small compared to the size of the rooms.

Microphone arrays are subject to the so-called *spatial aliasing effect* much like traditional sampling is subject to temporal aliasing. In order to avoid spatial aliasing, inter-microphone distance d_p of a pair p in a microphone array must be smaller than half-a-minimum wavelength. The signal frequency f_a above which spatial aliasing occurs depends on the microphone distance and the angle θ at which the acoustic signal incidents on the array. This is given as:

$$f_a = \frac{c}{2d_p cos\theta} \tag{3.4}$$

where c is the speed of sound (typically defined as $343m/s$).

In signal processing literature the angle of incidence is also known as Direction of arrival (DOA) and denotes the angle from which sound is arriving at the microphone array. Note that the DOA is typically measured in respect to the normal-to-the-array aperture. If we consider a single pair p from a microphone array, DOA can be schematically depicted as in Fig. 3.4.

Without loss of generality this considers $\mathbf{m_1}$ to be the location of the reference microphone. The signal will then arrive at the microphone at location $\mathbf{m_2}$ with a delay τ_p. The corresponding DOA angle θ is defined with respect to the broadside of the array and connected with any delay τ as:

$$\theta = \arcsin\left[\frac{\tau c}{f_s d_p}\right] \tag{3.5}$$

Thus, DOA is directly connected with the time delay between the microphones. Note that we restrict the estimation system to integer valued delays τ, for which several of the values of θ will correspond to the same integer delay. Thus, eq. (3.5) also defines the *resolution* of the array as a function of the values of d_p and f_s. The resolution of an array is a measure of the finite size of a point source as this is perceived by the array, and determines the closest distance to which two different source positions can be resolved.

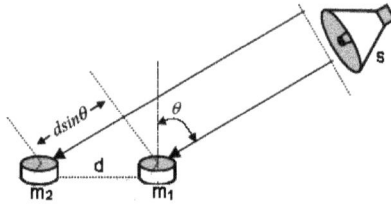

Fig. 3.4   Geometry of a single pair.

3.3 Audio trackers

3.3.1 *Linear intersection*

One apparent way to convert TDE measurements into a localisation measure is the conversion of the TDEs into corresponding DOAs (and respective
lines originating from each pair) and look for their crossing point. In the
presence of reverberation and noise the TDEs are subject to inaccuracies
and the corresponding lines cross in numerous points. The first breed of
trackers was purely deterministic since they attempted to filter out outlier
crossing points for every set of microphone data. This process was repeated
as more data became available. It us thus, of interest to examine such a
filtering mechanism.

If the TDE estimates provided by a set of microphone pairs are assumed
to deviate from the true TDE values due to additive, zero-mean, uncorre-
lated Gaussian noise, the Maximum Likelihood (ML) estimate of the source
location at frame t can be found through:

$$\mathbf{s}_t^{[ML]} = \arg\min_{\mathbf{s}}(\mathcal{J}(\mathbf{s})) \tag{3.6}$$

i.e. a minimisation of a least-squares error criterion [Kay (1993)] given as:

$$\mathcal{J}(\mathbf{s}) = \sum_{p=1}^{P} \frac{(\tau_p - \bar{\tau}_p(\mathbf{s}))^2}{\sigma_p^2} \tag{3.7}$$

where $\bar{\tau}_p(\mathbf{s})$ is the true TDE due to source at a potential location $\mathbf{s}$ and σ_p^2
is a variance estimate associated with each TDE τ_p. $\bar{\tau}_p(\mathbf{s})$ is a non-linear
function of $\mathbf{s}$. We can now deduce a closed-form solution for the source
location.

The Linear Intersection (LI) algorithm [Brandstein *et al.* (1997)] is a
simple solution to the above problem. Given a pair of microphones p, the

locus of potential source locations in space forms half a hyperboloid that for our purposes (far-field conditions) can be approximated by a cone with its vertex at the mid-point between the two microphones.

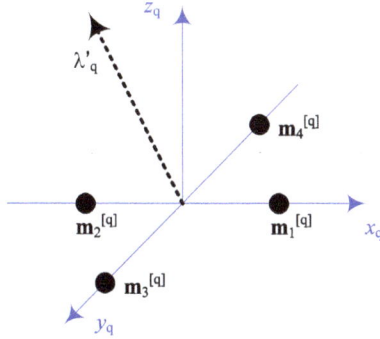

Fig. 3.5   Quadruple sensor arrangement and local Cartesian coordinate system.

Consider now two such pairs organised in a quadruple microphone array of planar form as in Fig. 3.5 (shown in the local coordinate system of the microphone array). The first pair has microphones at $\mathbf{m}_1^{[q]}$ and $\mathbf{m}_2^{[q]}$ while the second has them at $\mathbf{m}_3^{[q]}$ and $\mathbf{m}_4^{[q]}$ where $q = 1, \dots q_{max}$ refers to q_{max} different organisation of microphone pairs in quadruples. Thus, when the source is active at location $\mathbf{s}$, the first sensor pair determines a cone with direction angle equal to the corresponding estimated DOA ϕ as determined by eq. (3.5). The second pair specifies a similar cone at DOA χ. If the potential source location is restricted to the positive half space of the z-plane, the locus of the source locations common to the two formed cones is a line λ_q'. The remaining direction angle $\psi, 0 \leq \psi \leq \frac{\pi}{2}$ can be calculated by the law of cosines. For quadruple q, the bearing line λ_q' is given as:

$$\lambda_q' = u_q \mathbf{o}_q' \qquad (3.8)$$

where u_q is the range of a point on the line from the local origin $m_o^{[q]}$ (placed at the common mid-point of the two microphone pairs) and $\mathbf{o}_q' = [cos\phi_q, cos\chi_q, cos\psi_q]^T$ is the vector of the direction cosines for quadruple q. In order to express the bearing line in the global coordinate system we translate and rotate the local geometry in order to get:

$$\lambda_q = u_q \mathbf{o}_q + m_o^{[q]} \qquad (3.9)$$

where $\mathbf{o}_q$ represents the rotated direction cosine vector.

The LI approach considers the bearing lines created by a series of quadruples and estimates a number of potential source locations from the points of closest intersection for all pairs of corresponding bearing lines. The final source estimation is then given as a weighted average between those points.

If we consider the bearing lines λ_j, λ_k created by two quadruples the shortest distance between them is given as [Swokowski (1979)]:

$$d_{ij} = \frac{|(\mathbf{o}_j \times \mathbf{o}_k) \cdot (\mathbf{m}_j - \mathbf{m}_k)|}{|(\mathbf{o}_j \times \mathbf{o}_k)|} \tag{3.10}$$

Let $\hat{\mathbf{s}}_{jk}$ denote the point on λ_j with closest intersection to λ_k and correspondingly $\hat{\mathbf{s}}_{jk}$ the point on λ_k with closest intersection to λ_j. The linear intersection source estimate at frame t is then given as:

$$\mathbf{s}_t^{[LI]} = \frac{\sum_{j=1}^{q_{max}} \sum_{k=1, k \neq j}^{q_{max}} W_{jk} \hat{\mathbf{s}}_{jk}}{\sum_{j=1}^{q_{max}} \sum_{k=1, k \neq j}^{q_{max}} W_{jk}} \tag{3.11}$$

where W_{jk} is the weight associated with the potential source location $\hat{\mathbf{s}}_{jk}$ and given as:

$$\prod_{p=1}^{P} \mathcal{G}(\bar{\tau}_p(\mathbf{s}), \tau_p, \sigma_p^2) \tag{3.12}$$

where $\mathcal{G}(s, m, \sigma^2)$ is the evaluation of a Gaussian distribution with mean m and variance σ_p^2 at x. The use of the weighted average potentially filters out potential source estimates that are highly erroneous.

Even though the solution can prove effective for stationary sources in environments where reverberation and noise are not relatively high, the linear intersection variants are inappropriate for the environments we will examine i.e. enclosures with multiple moving humans and reverberation levels above average. Thus, for the remainder of the analysis we will concentrate on the state-space approach of PFs.

3.3.2   *Particle filtering*

Assuming a first order model for the acoustic-source dynamics the source state at any frame t is given as:

$$\boldsymbol{a}_t = [\mathcal{X}_t, \mathcal{Y}_t, \mathcal{Z}_t, \dot{\mathcal{X}}_t, \dot{\mathcal{Y}}_t, \dot{\mathcal{Z}}_t]^T \tag{3.13}$$

where $\mathbf{s}_t = [\mathcal{X}_t, \mathcal{Y}_t, \mathcal{Z}_t]$ is the current source location estimate and $[\dot{\mathcal{X}}_t, \dot{\mathcal{Y}}_t, \dot{\mathcal{Z}}_t]$ the corresponding source velocity. If we calculate the conditional probability density $q(a_t|\mathbf{y}_{1:t})$, we could then find the source location by choosing the state that is more likely given the sensor data until frame t. We can perform this by using the following relationship [Gordon *et al.* (1993)]:

$$q(a_t|\mathbf{y}_{1:t}) \propto q(\mathbf{y}_t|a_t)q(a_t|\mathbf{y}_{1:t-1}) \qquad (3.14)$$

where for clarity we have assumed $\mathbf{y}_t \equiv \mathbf{y}_{t:t}$. As discussed already, $q(\mathbf{y}_t|a_t)$ is called the likelihood function and expresses the means with which we value the states. $q(a_t|\mathbf{y}_{1:t-1})$ is known as the prediction density and it is given as [Gordon *et al.* (1993)]:

$$q(a_t|\mathbf{y}_{1:t-1}) = \int q(a_t|a_{t-1})q(a_{t-1}|\mathbf{y}_{1:t-1})da_{t-1} \qquad (3.15)$$

where $q(a_t|a_{t-1})$ is the state transition density, and $q(a_{t-1}|\mathbf{x}_{1:t-1})$ is the prior filtering density. The solution to (3.14) and (3.15) can be found using a Monte–Carlo simulation of a set of particles with associated discrete probability masses that estimate the source state [Arulampalam *et al.* (2002)]. For this we require a model of how the source propagates from a_{t-1} to a_t. To keep consistent with current literature we will use the Langevin model [Vermaak and Blake (2001)]. For the $\mathcal{X}_t$-coordinate this is defined as:

$$\dot{\mathcal{X}}_t = \alpha_{\mathcal{X}}\dot{\mathcal{X}}_{t-1} + \beta_{\mathcal{X}}G_{\mathcal{X}} \qquad (3.16)$$
$$\mathcal{X}_t = \mathcal{X}_{t-1} + \Delta T\dot{\mathcal{X}}_t \qquad (3.17)$$
$$\eta_{\mathcal{X}} = e^{-\beta_{\mathcal{X}}\Delta T} \qquad (3.18)$$
$$\beta_{\mathcal{X}} = v_{\mathcal{X}}\sqrt{1 - \eta_{\mathcal{X}}^2} \qquad (3.19)$$

where $G_{\mathcal{X}}$ is a normally distributed random variable, $\Delta T = L/f_s$ is the time separating two location estimates and f_s is the sampling frequency. Also, $v_{\mathcal{X}}$ refers to the steady-state velocity. Corresponding equations apply for $\mathcal{Y}_t$ and $\mathcal{Z}_t$.

Example 3.4. In MATLAB, a possible implementation of the propagation model for all three coordinates of a particle would be as follows:

```
ld = aX*ld + bX*randn(3,N);
l = l + DT*ld;
```

In the example `ld` represents the 3x1 velocity vector of each particle and `l` the 3x1 particle position. `aX` and `bX` represent $\alpha_{\mathcal{X}}$ and $\beta_{\mathcal{X}}$ respectively. ∎

We also need to decide on the likelihood functions that will operate on the microphone data. Depending on whether the system operates using TDE or BF measurements the likelihood function is different. These two likelihood functions will be presented in Sections 3.4.1 and 3.4.2 respectively.

The general structure of the proposed PF framework remains the same for both TDE and BF. It can be itemised as follows:

(1) Start with a set of particles $a_0^{[\iota]}$, $\iota = 1 \ldots N$ with uniform weights $w_0^{[\iota]}$, $\iota = 1 \ldots N$. For every new frame of data perform steps 2-8.

(2) Resample the particles from state $a_{t-1}^{[\iota]}$ using some resampling method (we used the *residual resampling* algorithm [Bolic *et al.* (2003)]) and form the resampled set of particles $\widetilde{a}_{t-1}^{[\iota]}$, $\iota = 1 \ldots N$.

(3) Using the Langevin model, propagate $\widetilde{a}_{t-1}^{[\iota]}$ to predict the current set of particles $a_t^{[\iota]}$.

(4) Take a set of frames of L samples from each microphone i.e. $\mathbf{x}_m^{[t]}$, $m = 1, 2, \ldots M$ and convert them into the frequency domain using an L-point Short Time Fourier Transfor (STFT) to get $\mathbf{X}_m^{[t]} = [X_m(\omega_0), X_m(\omega_1), \ldots, X_m(\omega_{L-1})]$, $m = 1, 2, \ldots M$. ω_l denotes the lth discrete frequency bin with $l = 0, 1, \ldots L - 1$.

(5) Using a localisation function convert the set of $\mathbf{X}_m^{[t]}$ into a localisation measurement i.e. a TDE or a BF measurement.

(6) Weight the particles using the likelihood function i.e. $w_t^{[\iota]} = q(\mathbf{y}_t | a_t^{[\iota]})$, $\iota = 1 \ldots N$ and normalise the weights so that they add up to unity.

(7) The source location for the current frame $\mathbf{s}_t$ is then given as the weighted average of the particles: $\mathbf{s}_t = \sum_{\iota=1}^{N} w_t^{[\iota]} \mathbf{l}_t^{[\iota]}$. In the last expression, $\mathbf{l}_t^{[\iota]}$ denotes the location vector of the ιth particle.

Example 3.5. Thus, in MATLAB, calculating the likelihood and retrieving a source estimate could be implemented as follows:

```
% Calculate likelihood for each particle:
w = ones(1,N);
for ii=1:N,
if ( l(1,ii)<MARGIN | l(1,ii)>(WIDTH-MARGIN) | ...
       l(2,ii)<MARGIN | l(2,ii)>(LENGTH-MARGIN) | ...
          l(3,ii)<MARGIN | l(3,ii)>(HEIGHT-MARGIN) ),
```

```
      w(ii) = 0; % Give zero weight if the
      %particle is outside the room.
   else
    for kk=1:NUMPAIRS
       Xpair = [X(pairs(kk,1),:);X(pairs(kk,2),:)];
       w(ii) = w(ii)*likelihood_F(l(:,ii), MIC(:,kk), c, Xpair);
                         % Likelihood for current particle.
      end
    end
end
NormFac = sum(w); % Normalisation factor
w = w/NormFac;   % Normalise weights

% Estimate of the source location based on current particles:
s_est(:,t) = sum([w; w; w].*1,3);
```

where in this case the function `likelihood_F` calculates the likelihood using TDE. In the case where we use BF, instead of looping through all pairs (in the inner loop, where we go up to the maximum number of pairs NUMPAIRS) we would have to loop through all microphones. w refers to the weights of the particles, MIC holds the coordinates of the microphones and pairs stores the desired pair organisation between the microphones. ∎

3.3.3 *Multiple speakers*

There are occasions where reverberation or noise sources can allow the particles to get trapped in a spurious location. Given this, we use the concept of an *external* likelihood evaluator (ELE) e_t that has the same architecture as the main PF a_t but it is initialised repeatedly at every frame t [Talantzis and Constantinides (2009)]. The particles of e_t are distributed randomly across the entire room. Thus, if these new particles estimate a source location that is d_e m away from the main particle filter for a significant amount of time T_e s then we can reset the locations of the particles of the main PF to those of the ELE. This also proves useful for scenarios where competing speakers are placed far apart. Nevertheless, at any instant the algorithm only tracks the location of the active speaker i.e. only estimates a single set of acoustic source coordinates.

We can thus introduce a further step to the algorithm to tackle the above scenario:

(8) If a sound source is active and the ELE $e_t^{[\iota]}$ returns a source estimate that remains at a distance greater than d_e m from the estimate of $a_t^{[\iota]}$ for more than T_e s then set $a_t^{[\iota]} = e_t^{[\iota]}$, $\iota = 1 \dots N$.

For the remainder of the Chapter we will refer to this modified tracking framework as the External Likelihood Evaluator partiCle TRAcker (ELECTRA).

3.4   Detection in audio

The first step in ASLT is the localisation process. This involves the detection of an acoustic event and the determination of whether this qualifies as a speech source. Detection in audio is performed using two main approaches:

(1) **TDE methods:** In this case the microphones of each microphone array are arranged in pairs. The different spatial locations of the microphones will result in corresponding recordings that are relatively delayed. TDE methods estimate the time difference between the two microphones of each pair by comparing the two recordings. In this case audio detection is performed by identifying the time delay that maximises some coherence metric between the microphones of each pair (it is assumed that coherence of the recordings due to noise will be low).

(2) **BF methods:** Detecting an audio source can also be performed using BF. In this case a multi-dimensional search over the vector space of possible source locations is performed in order to identify areas in the enclosure where power appears to be high (other measures can also be used as will be demonstrated in section 3.4.2).

Detection in audio using solely one of the above methods leads to ASLT systems with an inherent limitation. Coherency measures used for TDE and BF can prove inadequate when ASLT is used for tracking humans speaking. This is due to the fact that BF and TDE have no means of discerning between speech and an acoustic source emitting a signal that we do not wish to track e.g. a noise source exhibiting some level of coherency.

The solution to this problem involves the employment of a VAD system. In addition to discerning between noise and speech, a VAD module can potentially allow us to deal with the silent periods existing in the utterances of an active speaker.

3.4.1   *Time delay estimation*

In this case where we use TDE to detect audio the microphones are arranged in P pairs. Since the microphones of each pair p reside in different spatial locations, their corresponding recordings will be delayed with respect to each other by a relative time delay τ_p. TDE methods estimate the time difference between the two microphones of each pair p. Using all estimated τ_p the localiser can then provide an estimate of the current source location. Traditional systems typically do this by converting τ_p to a line along which the estimated source position is located. The problem of localisation could then reduce to finding the location which minimises the distance to each intersection points of the bearing lines. A characteristic example can be found in section 3.3.1.

Consider two microphones i, q belonging to the same pair p. Since the microphones reside in different spatial locations, their corresponding recordings will be delayed with respect to each other by a relative time delay τ_p when the source is active. Methods like the GCC [Knapp and Carter (1976)] functions (or one of its variants) exist for TDE. For any pair p the GCC-Phase Transform (GCC-PHAT) variant $R_t(\tau)$ is defined as the cross correlation of $\mathbf{x}_i^{[t]}$ and $\mathbf{x}_q^{[t]}$, filtered by a weighting function for a range of delays τ. In the frequency domain this is given as:

$$R_t(\tau) = \frac{1}{2\pi} \sum_{\omega_l} G(\omega_l) X_i(\omega_l) X_q^*(\omega_l) e^{j\omega_l \tau} \qquad (3.20)$$

with

$$G(\omega_l) = \frac{1}{|X_i(\omega_l) X_q^*(\omega_l)|} \qquad (3.21)$$

Ideally, $R_t(\tau)$ exhibits a global maximum at the lag value which corresponds to the correct τ i.e. at the delay corresponding to the correct DOA of the specific pair.

Example 3.6. In MATLAB calculating the GCC-PHAT function for a specific delay could be implemented as follows:

```
for noF = 1 : F

    C = xcorrPHAT(x1(noF,:), x2(noF,:));
    [maxC, idx] = max(abs(C));
    delay(noF) = (idx - ceil(length(C)/2)) ;

end
```

```
function [c] = xcorrPHAT(x,y)
M = length(x);
maxlag = M-1;
% Transform both vectors
X = fft(x,2^nextpow2(2*M-1));
Y = fft(y,2^nextpow2(2*M-1));
% Compute PHAT cross-correlation
c = ifft(X.*conj(Y)./(abs(X).*abs(Y)));
c = [c(end-maxlag+1:end) c(1:maxlag+1)];
```

where we are repeating the process for all F available frames of data originating from microphones in x1 and x2.   ∎

The GCC-PHAT algorithm is able to return accurate estimates of the relative delay when the environment is anechoic. However, it has a major drawback when used in an environment described by (3.1). In that case, reflections result in decreased system robustness since the peak provided by $R_t(\tau)$ may not always be the global maximum. This is particularly the case when T_{60} is not relatively low. In [Talantzis *et al.* (2005)] a new approach for TDE that utilises Mutual Information (MI) as a metric was presented.

We discuss the significance of the MI-based method in the frequency-domain. Since the analysis we will present is time-frame invariant we can drop t to express frame $\mathbf{X}_m^{[t]}$ as $\mathbf{X}_m$. The MI of two variables is an information theoretical measure that represents the difference between the measured joint entropy of the two variables (in our case these are the microphone signals of a pair p) and their joint entropy if they were independent. If there was no reverberation, the measurement of information contained in a frequency bin ω_l of $\mathbf{X}_i$ is only dependent on the information contained in $\mathbf{X}_q e^{j\omega_l \tau_p}$. In the case of the reverberant model though, information contained in a frequency bin ω_l of $\mathbf{X}_i$ is also contained in neighbouring bins of ω_l of $\mathbf{X}_q e^{j\omega_l \tau_p}$ due to the process of eq. (3.1). The same logical argument applies to the samples of $\mathbf{X}_q$. Thus, in order to estimate the information between the microphone signals, we use the marginal MI that considers jointly $\mathcal{N}$ neighbouring bins.

In a reverberant environment, microphone recordings are the superposition of attenuated and delayed versions of the original speech signal that exhibits a specific distribution. According to the Central Limit Theorem the resulting signal tends to have a normal distribution. Thus, if we assume that the microphone recordings exhibit normal distribution then for any set of frames in any pair p, the marginal MI between the two microphones is

[Cover and Thomas (1991)]:

$$I_{\mathcal{N}}(\tau) = -\frac{1}{2}\ln\frac{\det[\mathbf{C}(\tau)]}{\det[\mathbf{C}_{11}]\det[\mathbf{C}_{22}]} \qquad (3.22)$$

The joint covariance matrix $\mathbf{C}(\tau)$ is a concatenation of frames $\mathbf{X}_i$ and $\mathbf{X}_q$ shifted by different amounts:

$$\mathbf{C}(\tau) \approx$$

$$\mathbf{real} \left\{ \begin{bmatrix} \mathbf{X}_i \\ \mathcal{D}(\mathbf{X}_i, 1) \\ \vdots \\ \mathcal{D}(\mathbf{X}_i, N) \\ \mathcal{D}(\mathbf{X}_q, \tau f_s) \\ \mathcal{D}(\mathbf{X}_q, \tau f_s + 1) \\ \vdots \\ \mathcal{D}(\mathbf{X}_q, \tau f_s + \mathcal{N}) \end{bmatrix} \begin{bmatrix} \mathbf{X}_i \\ \mathcal{D}(\mathbf{X}_i, 1) \\ \vdots \\ \mathcal{D}(\mathbf{X}_i, N) \\ \mathcal{D}(\mathbf{X}_q, \tau f_s) \\ \mathcal{D}(\mathbf{X}_q, \tau f_s + 1) \\ \vdots \\ \mathcal{D}(\mathbf{X}_q, \tau f_s + \mathcal{N}) \end{bmatrix}^H \right\} \qquad (3.23)$$

$$= \begin{bmatrix} \mathbf{C}_{11} & \mathbf{C}_{12}(\tau) \\ \mathbf{C}_{21}(\tau) & \mathbf{C}_{22} \end{bmatrix}$$

where the **real**{.} operation returns only the real part of its argument. Function $\mathcal{D}(\mathbf{A}, n)$ shifts the frequency components contained in frame $\mathbf{A}$ by n samples. This is typically implemented by using an exponential with an appropriate complex argument.

If $\mathcal{N}$ is chosen to be greater than zero the elements of $\mathbf{C}(\tau)$ are themselves matrices. In fact for any value of τ, the size of $\mathbf{C}(\tau)$ is always $2(\mathcal{N}+1) \times 2(\mathcal{N}+1)$. We call $\mathcal{N}$ the *order* of the coherence function. $\mathcal{N}$ is really the parameter that controls the robustness of the MI-TDE function against reverberation. As noted, in the above equations and in order to estimate the information between the microphone signals, we use the marginal MI that considers jointly $\mathcal{N}$ neighbouring samples (thus the inclusion of delayed versions of the microphone signals). In this way the function of eq. (3.22) takes into account the spreading of information due to reverberation and returns more accurate estimates.

Example 3.7. In MATLAB calculating the MI function with order Nmax for a specific delay could be implemented as follows:

```
for noF = 1 : F
    X1 = fft(x1(noF,:));
```

```
    X2 = fft(x2(noF,:));

    for t=-shift:shift
        [IN(t+shift+1)] = real(mitde(X1, X2.*exp(j*2*pi*f*t/fs), order, fs));
    end
    IN = IN/log(exp(1));
    [dummy, Df(noF)] = max(IN);
    Df(noF) = Df(noF)-shift-1;
    clear IN;
end

function [IN] = mitde(X1_, X2_, Nmax, fs);
frame = length(X1_);
f = [0:fs/frame:(frame-1)*fs/frame];

for N = 0:Nmax
    X1(N+1,:) = X1_ .* exp(j*2*pi*f*N/fs);
    X2(N+1,:) = X2_ .* exp(j*2*pi*f*N/fs);
end
X = [X1;X2];
C = real(X*X');
IN = -0.5*log(det(C)/(det(C(1:N+1,1:N+1))*det(C(N+2:end,N+2:end))));
```

■

For the purposes of the ELECTRA framework we need to consider the following facts. At every frame t and after the microphones are organised in pairs, $I_\mathcal{N}$ is evaluated only at a set of candidate delays defined by the location of every particle ι. For two microphones i, q belonging in the same pair p the delay is given as:

$$\tau_p(a_t^{[\iota]}) = \frac{\left\|\mathbf{l}_t^{[\iota]} - \mathbf{m}_i\right\| - \left\|\mathbf{l}_t^{[\iota]} - \mathbf{m}_q\right\|}{c} \qquad (3.24)$$

where $\mathbf{m}_m$ denotes the location of the mth microphone at the pth pair and c the speed of sound (typically defined as $343m/s$). The $\|.\|$ operator denotes the Euclidean distance. Thus, in the ELECTRA framework when TDE is used the likelihood function for particle ι is given as:

$$q(\mathbf{y}_t|a_t^{[\iota]}) = \prod_{p=1}^{P} I_\mathcal{N}(\tau_p(a_t^{[\iota]})) \qquad (3.25)$$

The likelihood calculations when considering different TDE methods are similar. In the case of GCC-PHAT, $I_\mathcal{N}(\tau_p(a_t^{[\iota]}))$ is replaced by eq. (3.20). It must be stressed though that both of these expressions (or any other variant) do not necessarily represent properly normalised distribution functions but serve as such for the purposes of the algorithm.

3.4.2 *Beamforming*

Determining the source location using BF requires a single multi-dimensional search over the vector space of possible source locations. In most cases, the criterion to use involves looking for the location with maximum signal power [Bolic *et al.* (2002)]. Although direct methods do not require the calculation of intermediate time delays, a multi-dimensional search over source locations is required. Computationally this is potentially very demanding, especially for real-time implementations.

Beamforming exploits spatial diversity of desired speech and interfering speech or noise sources by combining multiple noisy input signals. A fixed beamformer combines the noisy signals of multiple microphones by a time-invariant filter-and-sum operation. The filtering operation can be designed to achieve constructive superposition towards a desired location. This is known as the delay-and-sum beamformer (DSB).

The output of a DSB beamformer steered to a location $\mathbf{l}$ is defined as:

$$Y(\mathbf{l}, \omega_l) = \frac{1}{M} \sum_{m=1}^{M} W_m(\mathbf{l}, \omega_l) X_m(\omega_l) \qquad (3.26)$$

where $W_m(\mathbf{l}, \omega_l)$ is the beamforming weight applied to the mth microphone. The DSB weights are given as:

$$W_m(\mathbf{l}, \omega_l) = e^{-j\omega_l c^{-1} \|\mathbf{l} - \mathbf{m}_m\|} \qquad (3.27)$$

Typically [Lehmann and Johansson (2007); Ward *et al.* (2003)], systems use the frequency-averaged output power of the BF to detect the true source location since one would expect the BF output to be large there. This is given as:

$$P(\mathbf{l}) = \sum_{\omega_l} |Y(\mathbf{l}, \omega_l)|^2 \qquad (3.28)$$

Example 3.8. Implementing a DSB in MATLAB that focuses at a given particle location l can be done as follows:

```
z = zeros(1,FFTlen); % DSB-based likelihood vector
for mic=1:NMIC,
ds = norm(l(:,n) - MIC(:,mic));        % delays to each microphone
W=exp(-j*2*pi*FREQFFT(1:FFTlen)*(ds/C)); % DSB weights
z = z + W.*X(mic,:);        % Sum over each delayed mic signal
end
out = 1/NMIC * sqrt(hamming(1:FFTlen)).*z; % DSB output for current particle
```

In this code we have also applied a Hamming window to the output using the `hamming` function. `FREQFFT` includes the discrete frequencies. ∎

Nevertheless, such power-based estimates are subject to the same problems as GCC in the case of TDE. Based on the same arguments presented in section 3.4.1 we propose a new BF approach that looks for the location at which the MI is maximum. The MI function is now given as:

$$I'_{\mathcal{N}}(1) = -\frac{1}{2} \ln \frac{\det[\mathbf{C}'(1)]}{\det[\mathbf{C}'_{11}]\det[\mathbf{C}'_{22}]} \qquad (3.29)$$

Given M sensors the covariance matrix is given as:

$$\mathbf{C}'(\tau) \approx$$

$$\text{real} \left\{ \begin{bmatrix} \mathbf{X}_1 \cdot \mathbf{W}_1(1) \\ \mathbf{X}_1 \cdot \mathbf{W}_1(1+\mathbf{b}_1) \\ \vdots \\ \mathbf{X}_1 \cdot \mathbf{W}_1(1+\mathbf{b}_{\mathcal{N}}) \\ \mathbf{X}_2 \cdot \mathbf{W}_2(1) \\ \mathbf{X}_2 \cdot \mathbf{W}_2(1+\mathbf{b}_1) \\ \vdots \\ \mathbf{X}_2 \cdot \mathbf{W}_2(1+\mathbf{b}_{\mathcal{N}}) \\ \vdots \\ \mathbf{X}_m \cdot \mathbf{W}_m(1) \\ \mathbf{X}_m \cdot \mathbf{W}_m(1+\mathbf{b}_1) \\ \vdots \\ \mathbf{X}_m \cdot \mathbf{W}_m(1+\mathbf{b}_{\mathcal{N}}) \end{bmatrix} \begin{bmatrix} \mathbf{X}_1 \cdot \mathbf{W}_1(1) \\ \mathbf{X}_1 \cdot \mathbf{W}_1(1+\mathbf{b}_1) \\ \vdots \\ \mathbf{X}_1 \cdot \mathbf{W}_1(1+\mathbf{b}_{\mathcal{N}}) \\ \mathbf{X}_2 \cdot \mathbf{W}_2(1) \\ \mathbf{X}_2 \cdot \mathbf{W}_2(1+\mathbf{b}_1) \\ \vdots \\ \mathbf{X}_2 \cdot \mathbf{W}_2(1+\mathbf{b}_{\mathcal{N}}) \\ \vdots \\ \mathbf{X}_m \cdot \mathbf{W}_m(1) \\ \mathbf{X}_m \cdot \mathbf{W}_m(1+\mathbf{b}_1) \\ \vdots \\ \mathbf{X}_m \cdot \mathbf{W}_m(1+\mathbf{b}_{\mathcal{N}}) \end{bmatrix}^H \right\} \qquad (3.30)$$

$$= \begin{bmatrix} \mathbf{C}'_{11} & \mathbf{C}'_{12}(\tau) \\ \mathbf{C}'_{21}(\tau) & \mathbf{C}'_{22} \end{bmatrix}$$

where $\mathbf{W}_m(1) = [W_m(1,\omega_0), W_m(1,\omega_1), \ldots, W_m(1,\omega_{L-1})]$ and the · operator denotes element-by-element multiplication. Also, $\mathbf{b}_{\mathcal{N}} = \frac{1}{\sqrt{3}}[c\mathcal{N}/f_s, c\mathcal{N}/f_s, c\mathcal{N}/f_s]$ denotes the distance travelled by sound in $\mathcal{N}/f_s$ seconds.

The MI-BF system is a direct analogy to the TDE system. To retain an analogy to the TDE system, we consider only positive integer delays i.e. instead of creating delayed versions of the frames (as we did in the

MI-TDE system of section 3.4.1) we use spatial delays by beamforming at corresponding locations.

At every frame t, $I'_\mathcal{N}$ is evaluated only at a set of candidate locations given by the positions of the particles. Thus, in the ELECTRA framework when BF is used the likelihood function for particle ι is given as:

$$q(\mathbf{y}_t|\boldsymbol{a}_t^{[\iota]}) = I'_\mathcal{N}(\mathbf{l}_t^{[\iota]}) \qquad (3.31)$$

The likelihood calculations when considering different BF methods are again similar. In the case of power-based BF, $I'_\mathcal{N}(\mathbf{l}_t^{[\iota]})$ is replaced by eq. (3.28). As with the TDE likelihoods these expressions and possible variants do not necessarily represent properly normalised distribution functions.

3.4.3   *Voice activity detection*

VAD systems attempt to replicate one of the basic operations of the human auditory system i.e. discriminate between voice and other ambient sounds. Most systems that work towards the solution of this problem use a single microphone approach and rely on extensive training prior to employment[Wrigley *et al.* (2005)]. The performance of all VAD systems relies again on the level of reverberation and noise.

The energy of a frame is the simplest criterion used to indicate the presence of speech [Deller *et al.* (1993)] in non-supervised systems. If the energy of a frame of audio data exhibits some threshold representing the maximum noise level then the system decides that speech is present. Other systems enhance the noisy speech before determining voice activity [Koh and Mortz (2000)]. Recently, the use of spectral entropy was introduced [Prasad *et al.* (2006)] as a measure and demonstrated that robustness increases when compared to energy methods. Information theoretic methods appear to be more robust than energy approaches because entropy exhibits insusceptibility to loudness variations in the recorded signals [Prasad *et al.* (2006)].

The VAD system proposed here uses more than one microphone to discern between a coherent acoustic source and spatially diffuse noise of low coherence. Measurement of coherency is performed using the MI functions of either the TDE or BF version of the tracker. Thus, no extra computational load is created if VAD is to be used along with the tracking system.

For the purposes of a tracking system VAD becomes significant when one considers the non-stationary nature of human speech that has a signif-

icant number of silence gaps between speech utterances. During these gaps
the PF framework would continue updating the source location with data
originating from an erroneous noise source. The algorithm would therefore
lose track of the actual sound source especially if during the silence gap the
source moves.

The scheme we present operates by first using the estimates of the MI, as
given by eq. (3.22) and eq. (3.29), when speech is absent. This represents a
threshold γ_t that is estimated from the particle with the highest likelihood.
Similarly to energy-based methods, the assumption is then that the speaker
is active when the MI is sufficiently higher than γ_t.

Threshold γ_t can be initialised by having access to some frames T_o where
the source is not active. The initial value of γ_t is then initialised to be:

$$\gamma_o = \frac{1}{T_o} \sum_{t=1}^{T_o} C_{\tilde{\iota}}^{[t]} \tag{3.32}$$

where $C_{\tilde{\iota}}^{[t]}$ is the coherence function used, calculated at the partcle $\tilde{\iota}$ with
maximum likelihood. Thus, for the MI-based systems $C_{\tilde{\iota}}^{[t]}$ is replaced with
the value of either eq. (3.22) or eq. (3.29) (depending on the tracking system
used) for time frame t. We further keep estimating γ_t for each of the later
frames in real-time. A fixed threshold would lead to decreased robustness
in varying acoustic environments. Thus, at frame t, γ_t adapts to a possible
environment change as:

$$\gamma_t = \frac{1}{2}(\mathcal{M}_0(\gamma_t, K) + C_{\tilde{\iota}}^{[t]}) \tag{3.33}$$

where $\mathcal{M}_0(\gamma_t, K)$ denotes the median value of the last K values of γ_t during
which no speech was detected. By using the median we avoid the effect of
sporadically large variations in the value of $C_{\tilde{\iota}}^{[t]}$ during frames that were
detected as silent.

A hangover scheme is another important consideration in a real-time
VAD system since it allows us to filter out false detections among frames
with the correct ones. Suppose for example that for a relatively long period
of time silence is being detected. The system then detects a single frame
where speech is present and immediately after a series of new frames without
speech. For the examined frame sizes this single speech detection should
be filtered out since speech utterances can not be that short. Thus, a
hangover scheme requires an initial state and two time constants δ_0 and δ_1

that determine after what amount of time we should switch to the no-speech or speech states respectively.

Fig. 3.6 shows the flow diagram of a hangover scheme used in the present work. Let's assume that initially the hangover scheme is at the no-speech state. In the steps to follow the algorithm receives a new frame of data and derives the presence or absence of speech i.e. $\alpha_t = 1$ or $\alpha_t = 0$ respectively. If $\alpha_t = 1$ then the system does not switch into this state immediately, rather it waits for δ_1 s during which more algorithm outputs are available. If these outputs persist during this period into $\alpha_t = 1$ then the state is changed. Similar logic then applies for switching back to $\alpha_t = 0$.

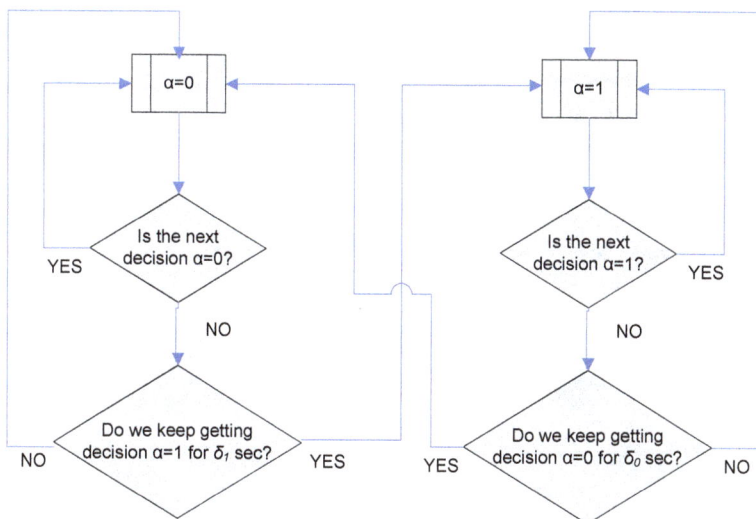

Fig. 3.6   Flow diagram of hangover scheme.

In general, $\delta_1 < \delta_0$ to allow for quick transition from no-speech to speech state and to restrict the opposite. This is in line with the fact that speech utterances are generally highly correlated with time [Prasad *et al.* (2006)]. This essentially means that since speech utterances are typically 100 *ms* or more in length the probability that a speech frame is followed by another one is high.

The integration of VAD in the tracking process is performed by not updating the source estimate when $\alpha_t = 0$ i.e. no-speech is detected. Additionally, during no-speech frames all particles are given uniform weights $w_t^{[\iota]}, \iota = 1 \ldots N$. Thus, during frames where speech is not present the geometrical variance of the particles will get larger (the particles will propagate randomly according to eqs. (3.16)–(3.19)). This way even if the speaker moves during the silent frames, he will most probably be in the vicinity of some particles when he begins speaking again.

Example 3.9. In MATLAB all of the above VAD funcionality can be implemented as follows:

```
for noF = 1 : F
    X=fft(xframe(:,:,noF).').'; % Compute the FFT of the current frame.
    for p=1:NUMPAIRS
        for t=-shift:shift
            % Compute the MI.
            IN(t+shift+1,p) = calcMI(X(pairs(p,1),:),...
              X(pairs(p,2),:).*exp(j*2*pi*f*t/fs), Nmax, fs);
        end

        [dummy, Df] = max(IN(:,p));
        tt(noF,p)=max(IN(:,p));
        if noF < K
            ttm(noF,p) = tt(noF,p);
        else
            ttm(noF,p) = median([tt(noF-4:noF-1,p); tt(noF,p)]);
        end

        vad_ = decide(ttm(noF,p));
        if vad_ == 0; v0 = v0 + 1; else v1 = v1 + 1; end
        vad(noF,p) = hangover(vad_, v0, v1)
    end
end
```

In this example we estimate speech activity **vad** at numerous pairs of microphones (defined by **NUMPAIRS**). **calcMI** returns an estimate of the MI function between two microphones (see example 3.7) while K identifies the length of the median filter. Function **hangover** simply tracks the repetition of 0s and 1s in order to decide on the final speech activity estimate. ∎

3.5   System parameters

3.5.1   *Performance measures*

There are two metrics that are used to evaluate the different systems. For their calculation we test the ASLT estimate at each time-frame against the ground truth (in the case of real experiments this was annotated manually), for the total duration of the test signals. The metrics are:

(1) The squared error for time frame t is given as:

$$\epsilon_t = \|\mathbf{s}_t - \bar{\mathbf{s}}_t\|^2 \tag{3.34}$$

where $\bar{\mathbf{s}}_t$ denotes the actual (manually annotated) source location. The first metric, the *Root Mean Square Error* (RMSE) is defined as the square root of the average value of ϵ_t over the total number of frames.

(2) The standard deviation for time frame t is given as:

$$\sigma_t = \sum_{\iota=1}^{N} w_t^{[\iota]} \left\| \mathbf{l}_t^{[\iota]} - \bar{\mathbf{s}}_t \right\|^2 \tag{3.35}$$

The second metric, the *Mean Standard Deviation* (MSD) is defined as the square root of the average value of σ_t over the total number of frames.

In the following sections, the above metrics are presented in metres. The lower their values, the better the performance of the system.

3.5.2 *Simulations*

In this section, we use simulated data to examine the performance of the PF framework for ASLT when the MI-based systems are used. The reader should treat this section as an indicative discussion of the ASLT problem rather than an exhaustive analysis of every possible ASLT scenario. For the remainder of the analysis we will refer to the MI-based systems as MI-TDE and MI-BF depending on whether the PF localisation measurement is performed with TDE or BF respectively. We compare the performance of MI-TDE and MI-BF to the systems that use GCC-PHAT and power-based BF. We will refer to these as GCC-PHAT and PBF. These systems have been evaluated in [Ward *et al.* (2003)]. The PF framework will remain identical for all algorithms.

During simulations comparison is performed using synthetic audio data at each sensor that was obtained using the image method for simulating small-room acoustics [Allen and Berkley (1979)]. The resulting filters are long enough to sample the reverberation time T_{60} i.e. they have $T_{60}f_s$ taps. The absorption coefficients were identical for all simulated walls. As presented in section 3.2.2, the image model assumes an omni-directional source model.

A single block of non-reverberant audio data has been used to obtain the results presented in this section. The source signal used was a speech utterance sampled at $f_s = 44100\ Hz$, pronounced by a male speaker and looped twice, yielding a total length of 120 s. The data received at each sensor was obtained by convolving this utterance with the corresponding impulse responses resulting from the image method between the positions of the source and the sensors. We calculate impulse responses and corresponding microphone signals for 30 random rotations and translations of the source trajectory. In all 30 cases though the positions of the microphones remain identical. The error metrics presented in the following paragraphs are the average value of the 30 simulations. By using this Monte–Carlo approach we filter out any outliers in terms of performance that might occur in simulations.

The ELECTRA PF framework uses $N = 50$ particles, $\beta_X = 10s^{-1}$ and $v_X = 1ms^{-1}$ (same values for the other coordinates).

The hangover scheme applied is kept identical for all algorithms with $\delta_0 = 0.74\ s$ and $\delta_1 = 0.37\ s$. Also, for each algorithm threshold γ_t is updated according to eq. (3.33) replacing $C_{\tilde{t}}^{[t]}$ with the corresponding coherence function for each algorithm. In the case of GCC-PHAT and PBF $C_{\tilde{t}}^{[t]}$ is replaced with eq. (3.20) and eq. (3.28) respectively and calculated for the same set of delays and locations.

The room dimensions were set to $5\ m \times 3.67\ m \times 2.58\ m$. An overview of the simulated room can be seen in Fig. 3.7. The figure also shows one of the translations of the acoustic source trajectory as well as the position of a second acoustic source that will be used for a set of the simulations. The microphones are organised in four microphone arrays (denoted as A', B', C', D') placed in an inverted-T topology as shown in the same figure. The pairs of microphones used for the TDE based systems can be seen in Table 3.1. The BF systems use all available microphones.

Table 3.1   Microphone pairs used for the simulations for TDE systems. Subscripts denote the microphone number of the corresponding array.

$Pair$	$Distance(cm)$
$A'_1 - A'_3$	30
$A'_2 - A'_3, A'_3 - A'_4$	20
$B'_1 - B'_3$	30
$B'_2 - B'_3, B'_3 - B'_4$	20
$C'_1 - C'_3$	30
$C'_2 - C'_3, C'_3 - C'_4$	20
$D'_1 - D'_3$	30
$D'_2 - D'_3, D'_3 - D'_4$	20

In all experiments unless otherwise stated the room has a reverberation time of 0.35 s, the SNR is 15 dB (random Gaussian noise was added to each microphone signal), $\mathcal{N} = 4$ and $L = 8192$ samples. The median filter used for the threshold calculation had a length of $K = 5$.

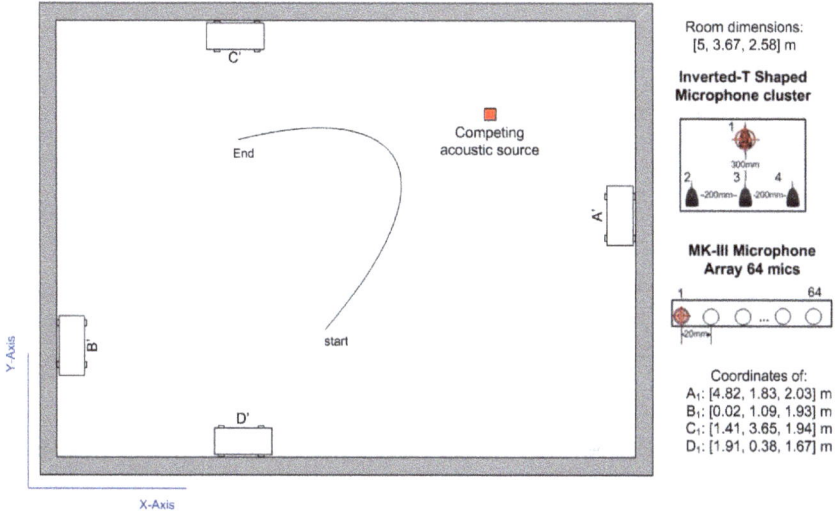

Fig. 3.7 Overview of simulated room.

3.5.2.1 *Effect of reverberation*

As noted in earlier sections, reverberation is typically measured by the reverberation time T_{60}. In order to quantify the effect of reverberation upon the ASLT systems, experiments were performed for a set of different environments distinguished by their reverberation times i.e. $T_{60} = [0, 0.10, 0.20, \ldots, 0.90, 1]$ s. Figures 3.8–3.9 show the effect of reverberation upon all algorithms. It can be observed that RMSE values get higher as the reverberation time increases. Nevertheless, the MI-based systems remain more robust than the other algorithms for all cases since the proposed coherence function of eq. (3.22) integrates means to filter out the effect of reverberation. For the used data set the MI-BF is the more robust system. The MSD values indicate if the source estimates are given from widely spread particles i.e. the smaller the value of the MSD metric the more certain the system is about the source location. In that respect, the MI-based systems perform better.

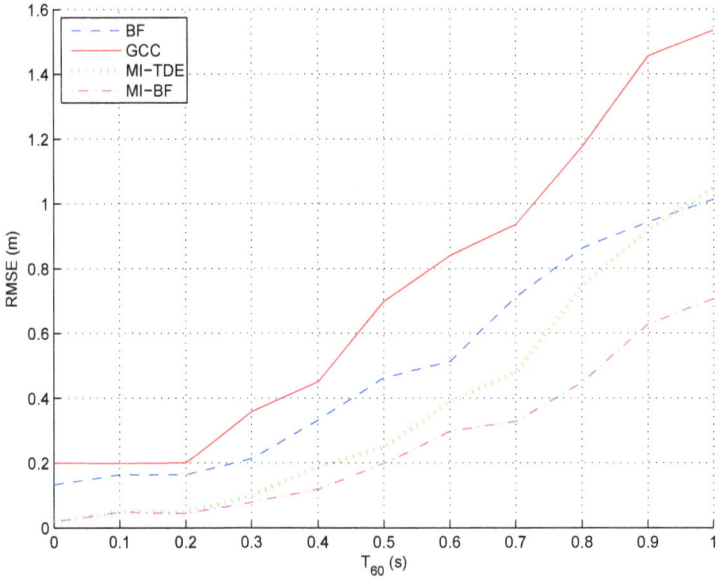

Fig. 3.8   Effect of reverberation time T_{60}. Average RMSE over 30 simulations is shown for all algorithms.

3.5.2.2   *Effect of the MI function*

Choosing the order $\mathcal{N}$ in the coherence functions of the MI-based systems affects the performance of the system. Figures 3.10(a)–3.10(b) show the RMSE for varying $\mathcal{N}$ for the MI-based systems and the way these compare to GCC and BF respectively. It appears that increasing the order N decreases the RMSE since the MI estimates and the resulting coherency calculations become more accurate. It is also worth noting that the performance of both MI-TDE and MI-BF reduce to the ones of GCC-PHAT and PBF for $\mathcal{N} = 0$ as the MI and covariance functions between two Gaussian variables are identical [Cover and Thomas (1991); Talantzis *et al.* (2005)].

3.5.2.3   *Effect of the VAD module*

Using a VAD system along with the ASLT allows the tracking module to focus on the dynamics model rather than the coherence functions during non-speech periods. In essence the scattering of particles increases, thus increasing the likelihood of some of them to stay close to the possibly moving source until it becomes active again. Figures 3.11(a)–3.11(b) show the

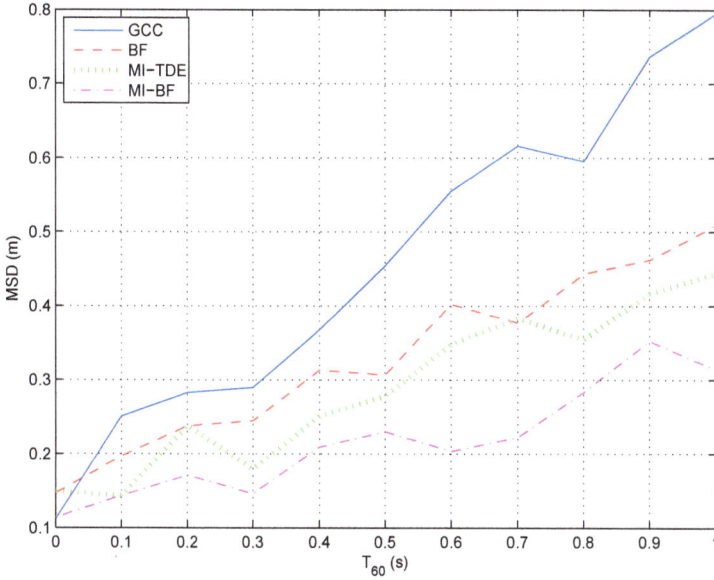

Fig. 3.9   Effect of reverberation time T_{60}. Average MSD over 30 simulations is shown for all algorithms.

degradation in performance when the VAD module is inactive. As noted, each algorithm uses a VAD module with a corresponding $C_i^{[t]}$ function. Performance of all systems is affected significantly.

3.5.2.4   *Effect of the array geometry*

As discussed earlier, the employment of microphones for ASLT is typically done by organising microphones in small clusters. In our case these clusters are represented by A', B', C', D' in Fig. 3.7. We look into three separate array topologies of four microphones located at Cartesian coordinates $\mathbf{m}_1, \mathbf{m}_2, \mathbf{m}_3, \mathbf{m}_4$. In order to make the topologies directly comparable we retain similar sizes. The three topologies used are T_1, T_2 and T_3 as these were presented in Fig. 3.3. The geometry of T_2 and T_3 (T_1 is presented in Table 3.1) and the pairs used for TDE systems are shown in Table 3.2.

Figures 3.12(a)–3.12(b) show that for the examined set of simulations T_2 appears to allow the ASLT system to produce more accurate estimates.

Table 3.2   Microphone pairs of T_2 and T_3 used for the simulations for TDE systems. Subscripts denote the microphone number of the corresponding array.

Pair	$Distance(cm)$ T_2	$Distance(cm)$ T_3
$A'_2 - A'_3, A'_3 - A'_4$	20	20
$A'_1 - A'_3$	30	30
$B'_2 - B'_3, B'_3 - B'_4$	20	20
$B'_1 - B'_3$	30	30
$C'_2 - C'_3, C'_3 - C'_4$	20	20
$C'_1 - C'_3$	30	30
$D'_2 - D'_3, D'_3 - D'_4$	20	20
$D'_1 - D'_3$	30	30

3.5.2.5   *Effect of interchanging sources*

During the employment of an ASLT system in a real room, interchanging sound sources create additional problems since they possibly reside far apart and become active in random time instances. A large distance between the two sources would affect the location accuracy as it would take longer for the particles to move from one source to the other. We simulate the effect of two sources that become interchangeably active by using the moving sound source used before and a second stationary sound source (as seen in Fig. 3.7). These become active one at a time at time instances $[20, 60, 90]$ s.

Figures 3.13–3.14 show the effect of employing an ELE as proposed in this work. In order to quantify the effect of ELE we use the Time Varying RMSE (TVRMSE) which represents the average error results, for every time instant, over the 30 simulations. Thus, the TVRMSE represents the changes of the RMSE in time. During the periods that a new source is active the TVRMSE increases rapidly. The systems that do not have an ELE slowly converge to the new source location. On the other hand the use of an ELE quickly copies the values of the primary PF to the newly detected source, thus saving time and decreasing the RMSE. Note that the error peaks drop as the two sources come closer. It can also be seen that the BF systems appear to be more responsive in converging to the new location than the TDE ones while also exhibiting a lower MSD which is denoted by the overall variation of the RMSE.

In all cases the MI-based systems perform better. The MI-BF system has the lower error rates among all systems tested since it requires approximately 2 s to converge to the new source location. Without the ELE this rises to 4–5 s.

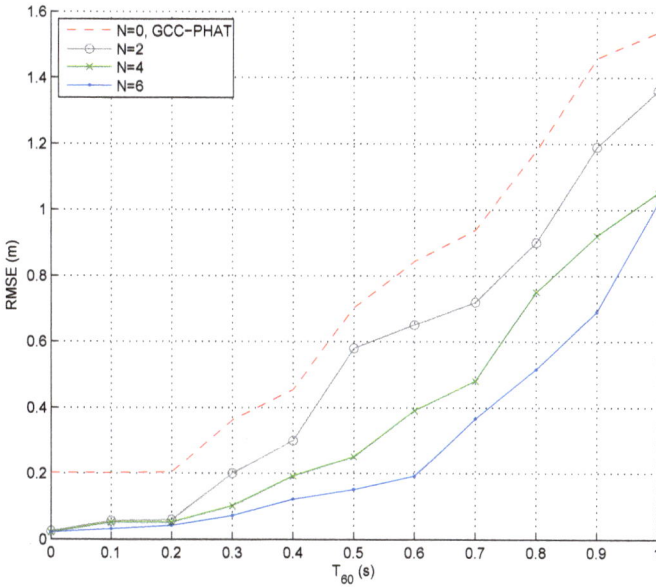

(a) MI-TDE algorithm.

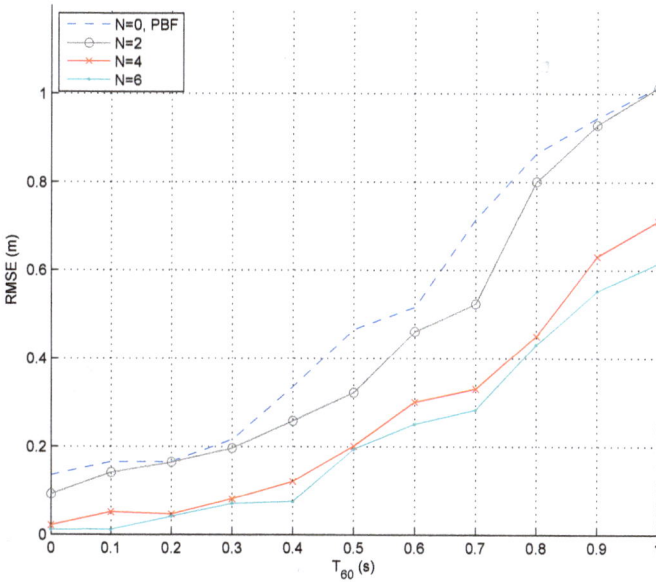

(b) MI-BF algorithm.

Fig. 3.10 Effect of order $\mathcal{N}$. Average RMSE over 30 simulations.

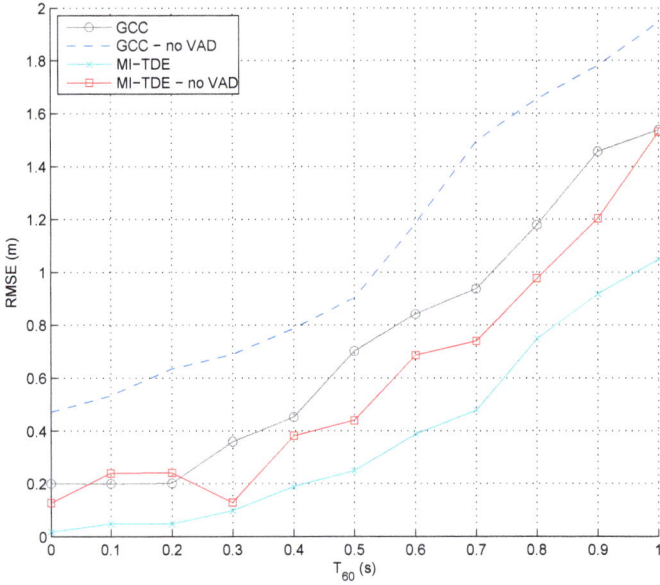

(a) TDE algorithms.

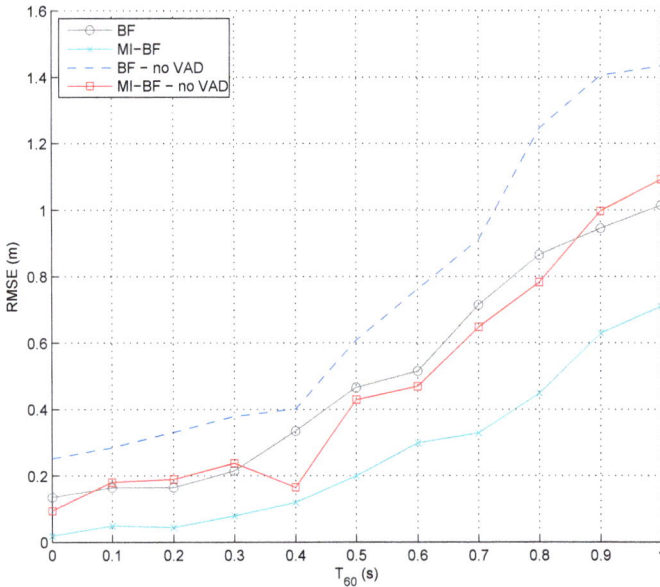

(b) BF algorithms.

Fig. 3.11   Effect of the VAD module. Average RMSE over 30 simulations.

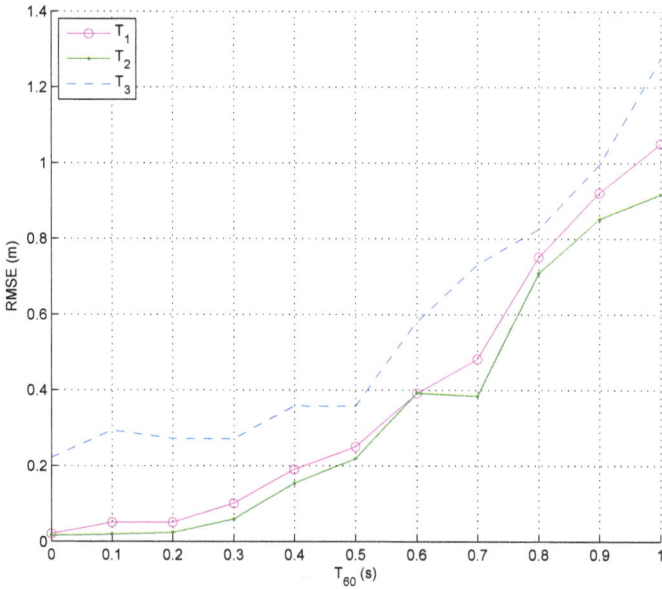

(a) TDE algorithms.

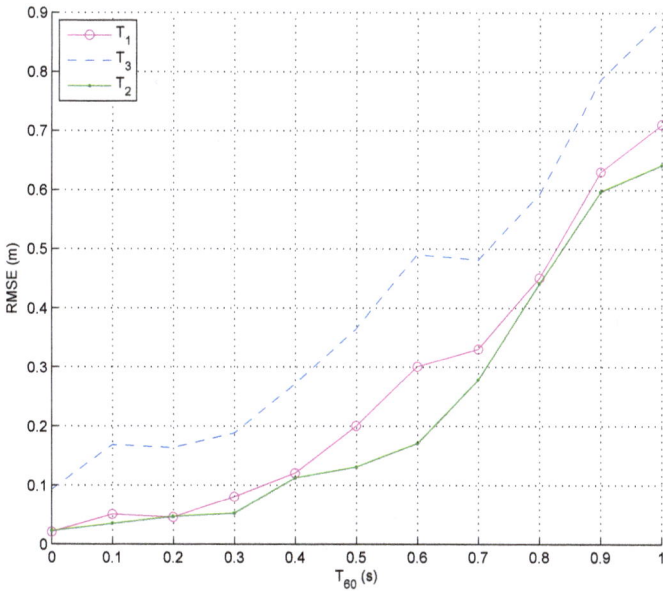

(b) BF algorithms.

Fig. 3.12   Effect of the array topology. Average RMSE over 30 simulations.

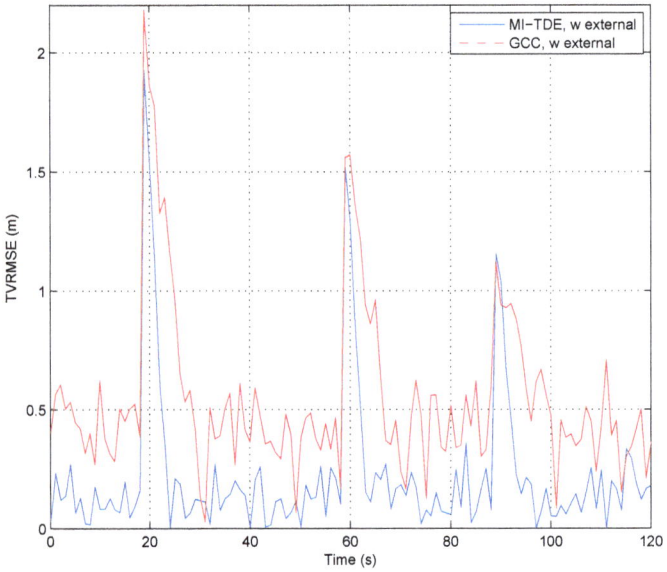

(a) With ELE.

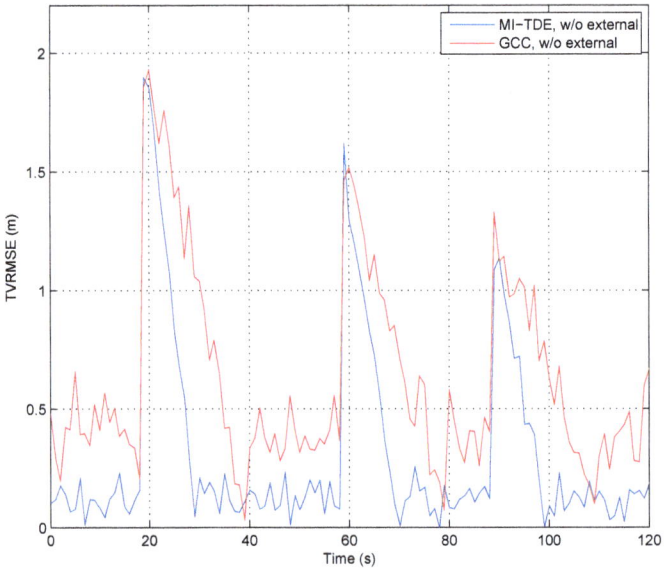

(b) Without ELE.

Fig. 3.13   Effect of interchanging sources. TVRMSE over 30 simulations is shown for the MI-TDE and GCC algorithms.

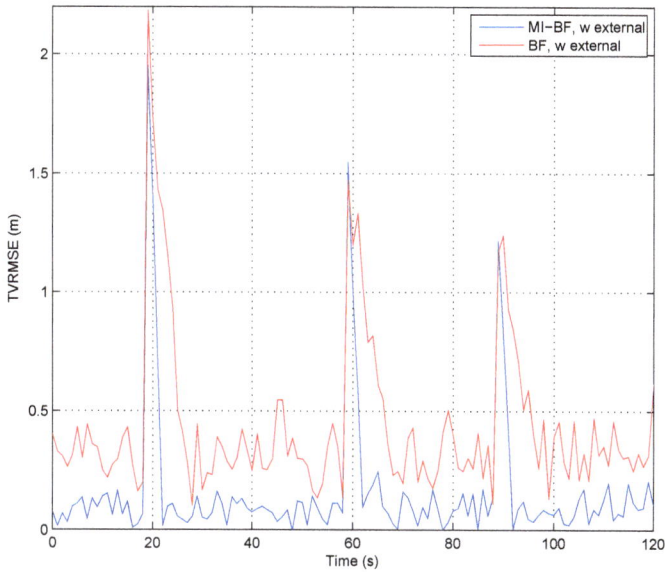

(a) With ELE.

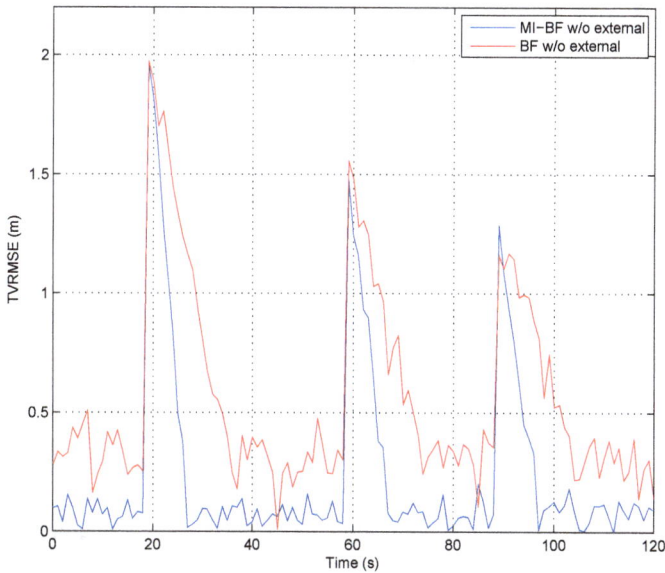

(b) Without ELE.

Fig. 3.14 Effect of interchanging sources. TVRMSE over 30 simulations is shown for the MI-BF and BF algorithms.

Chapter 4

Visual Tracking

4.1 Introduction

Visual tracking refers to estimating across time the state of targets of interest given the feed of cameras, i.e. successive video frames where these targets are depicted. Any of the tracking frameworks of Chapter 2 can be used, utilising evidence collected from the video frames to support the tracker hypotheses.

Depending on the deployed cameras, a visual tracker can operate on one or multiple synchronised camera views. The evidence is collected from a portion of each of these images which is dictated by the state. The mapping of the state onto the image(s) depends on two factors:

- The nature of the state. The state can involve pixel coordinates on the image, or three-dimensional (3D) world coordinates. In the latter case, a mapping from the 3D world coordinate system to the image plane(s) is needed. This is accomplished using the intrinsic and extrinsic camera parameters obtained by camera calibration [Zhang (2000)], as detailed in section 4.2.
- The nature of the evidence collected from the images. Visual person tracking differs a lot from radar tracking [Blackman (1986); Blackman and Popoli (1999)] in the sense that targets depicted in images do not generate a signal to be discriminated from the noise of the background clutter by the traditional means of signal-to-noise ratio. For example, in visual person tracking, pixels belonging to people can have almost any colour, especially in terms of clothing. The same is true for pixels belonging to background. Thus asking for some particular signal level may not be helpful. The different types of evidence collected are detailed in sections 4.4–4.8.

All the measurement cues can be used to provide the different visual tracking algorithms with evidence. In the final sections of this chapter, the particulars of deterministic (section 4.10) and stochastic (section 4.11) trackers are discussed, and example systems are detailed.

4.2   From 3D world to 2D images

In order not to restrict the operation of the trackers on the image plane, and to be able to use multiple images from synchronised cameras, one needs to associate the 3D world to the image plane(s). A camera views the surrounding 3D space and projects it onto the image plane. The geometry of this projection depends on two factors, the orientation of the camera and the lens.

The orientation of the camera relative to the world can be described using two 3D Cartesian coordinate systems: that of the world and that of the camera (see Fig. 4.1). The world coordinate system is usually defined with the ground being the x-y plane (especially for flat surfaces) and the height being measured along the z axis. The origin is positioned at a convenient point, e.g. a corner of a room. The camera coordinate system has its origin at the centre of projection of the camera, and its z_c axis along the principal ray of the camera [Shapiro and Stockman (2001); Forsyth and Ponce (2002)]. The orientation of the other two axes approximately coincides with that of the two image plane axes, apart from a possible skew considered later in this section. Its offset from the world coordinate system is represented by the vector $\mathbf{T}_c$, termed translation vector. Its axes are also oriented differently than those of the world coordinate system. They have to be rotated to match. This is achieved by the rotation matrix $\mathbf{R}_c$. The coordinates of a point in the world coordinate system $\mathbf{P}$ are hence related to those in the camera coordinate system $\mathbf{P}_c$ by:

$$\mathbf{P} = \mathbf{R}_c \mathbf{P}_c + \mathbf{T}_c \qquad (4.1)$$

The rotation matrix and translation vectors are the extrinsic parameters of the camera.

The camera coordinates of a point are not the coordinates on the image plane. It is the lens type and the way it is mounted on the camera that govern the way a point in the camera coordinate system is projected onto the image plane. The first part of this projection involves the non-linear distortion of the camera coordinates. This non-linear distortion has two components [Zhang (2000)]. The first component is radial, i.e. a function of

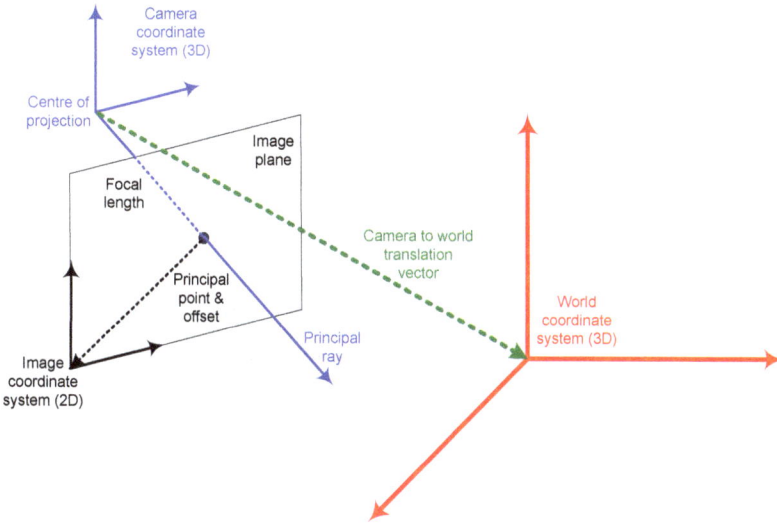

Fig. 4.1 Viewing a scene with a camera to project it onto an image plane. Three coordinate systems are involved.

(even) powers of the distance from the principal point. A linear combination of the second ($k_r^{(2)}$), fourth ($k_r^{(4)}$) and sixth ($k_r^{(6)}$) powers of the distance are used. The second distortion component is tangential and is determined by the two coefficients $k_t^{(1)}$ and $k_t^{(2)}$ as shown below.

Each line connecting the origin of the camera coordinate system (centre of projection) and a point on that system is projected onto the same point in the image plane, giving rise to the depth uncertainty of single camera imaging. For this reason the correspondence between the depth-normalised coordinates of the camera coordinate system and the image plane coordinates is sought. Starting from the camera coordinates $[x_c, y_c, z_c]^T$, the depth-normalised camera coordinates are:

$$\mathbf{x}_n \equiv \begin{bmatrix} x_n \\ y_n \end{bmatrix} = \begin{bmatrix} x_c/z_c \\ y_c/z_c \end{bmatrix} \tag{4.2}$$

Setting $r^2 \equiv x_n^2 + y_n^2$, the (radial and tangential) distorted coordinates $\mathbf{x}_d$ are [Zhang (2000); Bouguet (2008)]:

$$\mathbf{x}_d \equiv \begin{bmatrix} x_d \\ y_d \end{bmatrix} = \mathbf{x}_r + \mathbf{x}_t \tag{4.3}$$

where the radial distortion is modelled using three coefficients $k_r^{(i)}$ of order

i equal to two, four and six:

$$\mathbf{x}_r = \left(1 + k_r^{(2)} r^2 + k_r^{(4)} r^4 + k_r^{(6)} r^6\right) \mathbf{x}_n \qquad (4.4)$$

and the tangential distortion is modelled using two coefficients $k_t^{(i)}$, $i = 1, 2$:

$$\mathbf{x}_t = \begin{bmatrix} 2k_t^{(1)} x_n y_n + k_t^{(2)} \left(r^2 + 2x_n^2\right) \\ k_t^{(1)} \left(r^2 + 2y_n^2\right) + 2k_t^{(2)} x_n y_n \end{bmatrix} \qquad (4.5)$$

The second part of the camera coordinate system projection onto the image plane involves the linear projection of $\mathbf{x}_d$. It is determined by the following parameters, taken from the pinhole camera model:

- Focal length: This is the distance of the principal point from the centre of projection.
- Non-square pixels: The aspect ratio of the pixels together with the focal length result in the scaling (different along each axis) of the viewed objects. Together they are denoted by the focal length 2×1 vector $\mathbf{f}_c = [f_x, f_y]^T$.
- Principal point offset: This is the 2×1 translation vector $\mathbf{c}_c = [c_x, c_y]^T$ between the principal point on the image plane and the origin of the pixel coordinates.
- Skew: The skew coefficient α_c accounts for the fact that the camera coordinate system axes x_c and y_c are only approximately parallel to the axes of the image coordinate system x_c and y_c respectively. In most cases skew is approximated with zero.

Thus the pixel coordinates of the world point are a function of the distorted coordinates $\mathbf{x}_d$, given by:

$$\mathbf{x}_p = \begin{bmatrix} f_x \cdot (x_d + \alpha_c y_d) \\ f_y \cdot y_d \end{bmatrix} + \mathbf{c}_c \qquad (4.6)$$

The distortion coefficients and the pinhole model parameters form the intrinsic camera parameters. Given these intrinsic camera parameters, eqs. (4.1) to (4.6) map any point in the 3D world to a pixel in the image plane of the camera. Note that the pixel coordinates might not actually be contained within the image boundaries, if the particular 3D point is not viewed by the camera. An example of the mapping with the MATLAB implementation of equations can be found in example 4.1.

Example 4.1. Project points of the 3D world coordinate system onto image planes.

The target camera views are stored in the four-dimensional array of images I. The camera parameters for each of them are stored in the array of structures camData. Both can be found in the MAT-file UPCoutlineData.mat.

The camera calibration parameters are stored in MATLAB variables following the notation of the Camera Calibration Toolbox for MATLAB [Bouguet (2008)]. The extrinsic parameters are denoted by T and R (translation vector and rotation matrix respectively). The intrinsic parameters for the linear transformation are denoted f (focal length vector), c (principal point) and alpha (skew coefficient). As for the five non-linear distortion parameters, they are stored in the 5×1 vector kc, ordered as $\left[k_r^{(2)}, k_r^{(4)}, k_t^{(1)}, k_t^{(2)}, k_r^{(6)}\right]^T$. Note that the sixth-order radial distortion coefficient $k_r^{(6)}$ comes fifth in k. For compactness in notation, all these parameters are grouped in the given structure camData.

The MATLAB function world2cam() implements the projection from the n points with world coordinates in the 3-by-n matrix P to the image plane coordinates in the 2-by-n matrix xp. To do so eqs. (4.1) to (4.6) are implemented. The comments in the code identify which equation is implemented by each block of code.

```
function xp=world2cam(P,R,T,f,c,k,alpha)
n = size(P,2);
% Camera coordinate system (4.1)
Pc = R'*(P - repmat(T,[1 n]));
% Depth normalisation (4.2)
inv_Z = 1./Pc(3,:);
xn = (Pc(1:2,:) .* (ones(2,1) * inv_Z)) ;
% Radial distortion (4.4):
r2 = xn(1,:).^2 + xn(2,:).^2;
r4 = r2.^2;
r6 = r2.^3;
xr = xn .* repmat(1 + k(1) * r2 + k(2) * r4 + k(5) * r6, 2,1);
% Tangential distortion (4.5):
a1 = 2.*xn(1,:).*xn(2,:);
a2 = r2 + 2*xn(1,:).^2;
a3 = r2 + 2*xn(2,:).^2;
xt = [k(3)*a1 + k(4)*a2 ; k(3) * a3 + k(4)*a1];
% Add distortion (4.3):
xd = xr + xt;
% Pixel coordinates (4.6):
xp = [xd(1,:) + alpha*xd(2,:);xd(2,:)] .* repmat(f,1,n)  +  repmat(c,1,n);
```

Fig. 4.2   Projection of the head centres onto the five camera views for example 4.1.

The five different camera views are shown in Fig. 4.2. If coordinates for the heads of the four people visible in the scene are given in the matrix P, then the following code utilises `world2cam()` to project the four head centres onto all five camera views.

```
load UPCoutlineData
P=[680 1390 1680;...
   3420 2600 1720;...
   1870 3810 1580;...
   3060 3720 1680]';
for cam=1:5
    figure(cam),imshow(I(:,:,:,cam),'Border','Tight')
    xp=world2cam(P,camData(cam).Rc,camData(cam).Tc,camData(cam).fc,...
                camData(cam).cc,camData(cam).kc,camData(cam).alpha_c);
    hold on,plot(xp(1,:),xp(2,:),'r.', 'markersize',24),hold off
end
```

■

Mapping of real-world points onto images is very useful for 3D tracking. In the model-based approach for 3D tracking [Bernardin *et al.* (2009)], a 3D model is evaluated and projected onto the images from the different cameras, to collect evidence of the target. Model-based 3D tracking evidence collection is further discussed for the case of outline projection in example 4.10, of section 4.6.2.

The intrinsic and extrinsic parameters of the camera/lens combination are obtained by the process of camera calibration. This process involves the processing of images depicting various views of a co-planar chequerboard pattern by camera calibration software, such as the Camera Calibration Toolbox for MATLAB from the California Institute of Technology [Bouguet

(2008)] or the Camera Calibration Toolbox for IDL from the Institute of Robotics and Mechatronics [Strobl *et al.* (2005)].

4.3   3D imaging

In the previous section the forward problem of mapping a real-world point into the camera planes has been considered. The inverse problem is also interesting: map an image point into a real-world point. There are two difficulties in attempting to do so:

- The forward problem is non-linear, hence its analytic inversion is not possible. So the inverse problem is approximated numerically.
- Even the numerical approximation does not lead to a single point on the real-world coordinate system due to depth uncertainty. Recall the depth normalisation of (4.2), which cannot be reversed, since depth information does not exist after it.

Depth uncertainty is only rarely resolved using a single camera. Only when the projection of the object into the image plane has known dimensions can the depth be estimated, as discussed in section 4.8.2 concerning faces. On the other hand, depth uncertainty is readily resolved using multiple cameras. Referring to Fig. 4.3 for a two camera example, a real-world object at point $\mathbf{x}_{3D}$ (in the world coordinate system) is depicted at pixels $\mathbf{x}_p^{(1)}$ and $\mathbf{x}_p^{(2)}$ in the two image planes. These pixels correspond to depth-normalised camera coordinates $\mathbf{x}_n^{(1)}$ and $\mathbf{x}_p^{(2)}$ respectively, that define two depth uncertainty lines. These lines are expressed in the world coordinate system by means of the extrinsic parameters of the respective cameras, i.e. the translation vectors $\mathbf{T}_1$ and $\mathbf{T}_2$, as well as the rotation matrices $\mathbf{R}_1$ and $\mathbf{R}_2$. Ideally the depth uncertainty lines would intersect at $\mathbf{x}_{3D}$, but this is rarely the case: unless the object is a sphere, the centre of its projection on an image plane does not correspond to the centre of the object. Localisation errors on the image planes make this more evident. Hence we actually seek the point of closest intersection, located midway along the segment of minimum distance between the two lines.

Generalising to C cameras viewing the object of interest, we first approximate $\mathbf{x}_n^{(c)}$ from $\mathbf{x}_p^{(c)}$, $c = 1, \ldots, C$. As already mentioned this inverse problem is solved only numerically, due to the fact that eqs. (4.3)–(4.5) are non-invertible. This problem is considered by the Calibration Toolbox for MATLAB [Bouguet (2008)], and is addressed by the functions `normalize()`

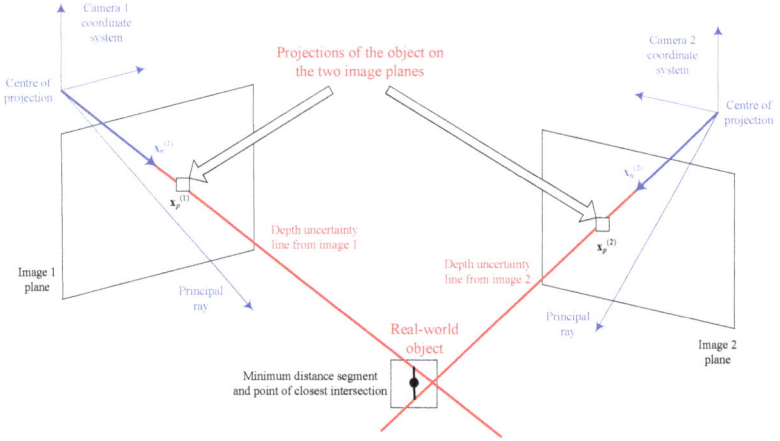

Fig. 4.3 Viewing a scene with a camera to project it onto an image plane. Three coordinate systems are involved.

and `comp_distortion_oulu()`. The former undoes the linear part in (4.6), yielding $\mathbf{x}_d^{(c)}$, and calls the latter to approximate $\mathbf{x}_n^{(c)}$.

Having the $\mathbf{x}_n^{(c)}$ that represent the depth uncertainty lines, the point of closest intersection $\mathbf{x}_{3D}$ is obtained by solving the following set of equations:

$$
\begin{bmatrix}
\mathbf{I}_3 & -\mathbf{R}_1\mathbf{x}_n^{(1)} & \mathbf{0}_3 & \cdots & \mathbf{0}_3 \\
\mathbf{I}_3 & \mathbf{0}_3 & -\mathbf{R}_2\mathbf{x}_n^{(2)} & \cdots & \mathbf{0}_3 \\
\vdots & \vdots & \vdots & \ddots & \vdots \\
\mathbf{I}_3 & \mathbf{0}_3 & \mathbf{0}_3 & \cdots & -\mathbf{R}_C\mathbf{x}_n^{(C)}
\end{bmatrix} \cdot \mathbf{x}_{3D} =
\begin{bmatrix}
\mathbf{T}_1 \\
\mathbf{T}_2 \\
\vdots \\
\mathbf{T}_C
\end{bmatrix}
\tag{4.7}
$$

where $\mathbf{I}_3$ is the 3×3 identity matrix and $\mathbf{0}_3$ is the 3×1 zero vector. This is an overdetermined system of $3 \times C$ equations and three unknowns. It is readily solved using the pseudo-inverse of the left-hand matrix. Note that prior to multiplying with the $\mathbf{R}^{(c)}$, the $\mathbf{x}_n^{(c)}$ are extended to 3D by appending unity as the normalised third dimension (depth). It is not necessary that all of the C cameras are viewing the object of interest. In this case the cameras not viewing the object are omitted from the set of equations.

The implementation and solution of the system in (4.7) is demonstrated in example 4.2.

Example 4.2. The following MATLAB function `locate3D()` accepts the pixel coordinates in the $2 \times C$ matrix `Xp` of an object in the C images and the camera parameters (intrinsic and extrinsic) in the array of C structures cp. It returns the 3D world coordinates x3D of the object. When an object is not viewed by a camera whose parameters are passed in cp, its pixel coordinates are passed as NaN.

The first loop finds the normalised coordinates per camera, extends them to 3D and multiplies by the rotation matrix. It also appends the translation vector into a super-vector Tc and counts the number of cameras C that view the object.

The second loop creates the left-hand matrix of (4.7), storing it in A. The system is then solved for the 3D coordinates employing the pseudo-inverse of A, implemented using function pinv().

```
function x3D=locate3D(Xp,cp)
Tc=[];Xn=[];
C=0;
for c=1:size(Xp,2)
    if ~isnan(Xp(1,c))
        xn=normalize(Xp(:,c),cp(c).fc,cp(c).cc,cp(c).kc,cp(c).alpha_c);
        Xn=[Xn;cp(c).Rc*[xn;1]];
        Tc=[Tc;cp(c).Tc];
        C=C+1;
    end
end
A=[];
% At least two cameras are viewing the object
if C>=2
    % Left-hand matrix in (4.7)
    for c=1:C
        Ac=[eye(3) zeros(3,C)];
        Ac(:,3+c)=-Xn((c-1)*3+1:3*c);
        A=[A;Ac];
    end
    x3D=pinv(A)*Tc;
    x3D=x3D(1:3);
else
    x3D=[NaN;NaN;NaN];
end
```

■

The mapping from multiple camera views into the 3D space is the cornerstone of data-driven 3D tracking. According to this tracking paradigm, 2D trackers operate independently on separate camera views; then the 2D tracks belonging to the same target are collected into a 3D one [Zhang et al. (2007); Tyagi et al. (2007); Katsarakis et al. (2008)]. The data-driven approach to 3D tracking is depicted in Fig. 4.4.

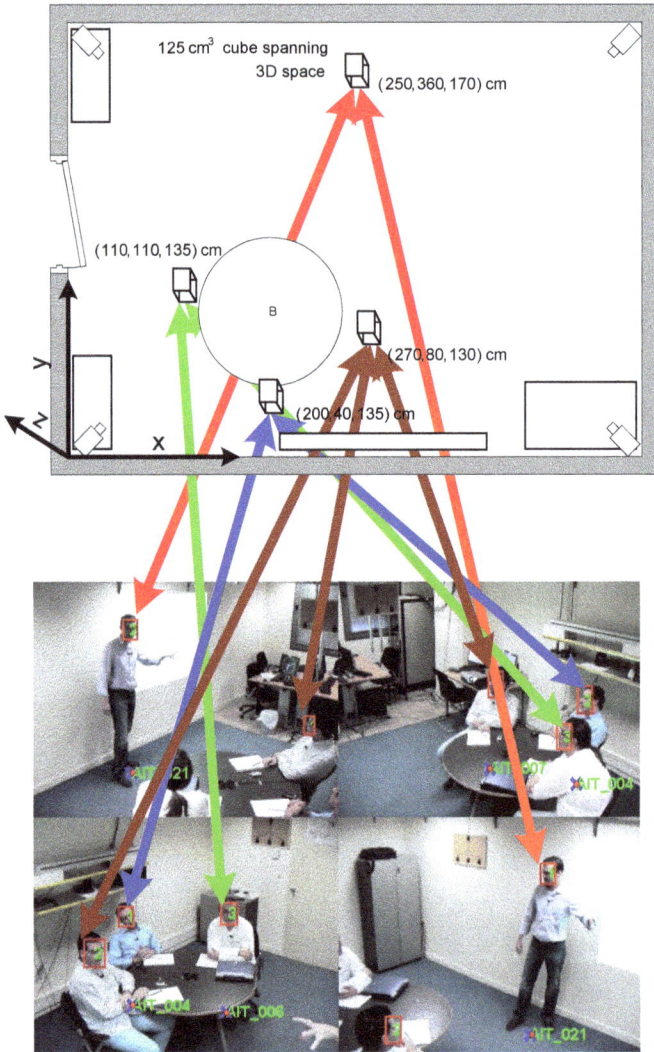

Fig. 4.4   Data-driven 3D tracking by combination of the independent 2D face tracks (red rectangles) from multiple views that correspond to the same target. After combination, the 3D coordinates obtained using `locate3D()` are projected onto the floor (height is ignored) and are shown as red "x" marks.

4.4 Measurements in images

States $\mathbf{x}$ involving either image plane coordinates or 3D world coordinates are both mapped onto regions $R_{\mathbf{x}}$ of the available image plane(s). In its simplest form $R_{\mathbf{x}}$ is a rectangle. For example, the state vector can be $\mathbf{x} = [x, y, a]^T$, where x and y are the image coordinates of the top-left corner of the target on the image plane, and a is the scale factor relative to a reference rectangle of width W and height H. $R_{\mathbf{x}}$ is then a rectangle with top-left corner pixel coordinates (x, y) and bottom-right corner pixel coordinates $(x + aW - 1, y + aH - 1)$. There is no need to constraint $R_{\mathbf{x}}$ to being a rectangle. It can be the interior of an arbitrary shaped polygon, such as the outline of a human body. However complicated the shape of $R_{\mathbf{x}}$ and the form of the state $\mathbf{x}$ are, the mapping $\mathbf{x} \rightarrow R_{\mathbf{x}}$ always exists.

It is within the regions $R_{\mathbf{x}}$ that visual evidence is collected to support the tracing hypotheses by means of measuring image properties. The different properties correspond to visual cues such as colour, texture and shape. There are many different cues to measure in the video frames, each offering advantages but suffering from disadvantages. A perfect cue would be:

- Discriminative from the background, to reduce ambiguity.
- Robust to the ways the target can be depicted. The image of the target can change due to many factors. Illumination changes as the target moves relative to the light sources, or the sources themselves change. Pose changes as the target rotates relative to the cameras. Finally, shape changes as the target flexes (or in the case of human faces, changes expression).
- Persistent, i.e. detectable at most of the frames at which the target is visible.

Unfortunately there is no single cue that has all these qualities: colour is persistent and quite robust to the appearance of the target, but may not be discriminative from the background. Target outline is discriminative and persistent (when the outline model describes the object adequately) but target flexing can render learning an outline model very difficult, and applying the model very complicated. Motion is again discriminative, especially in fixed camera setups, but is not persistent as the target can spend long periods almost immobile. Finally, dedicated target detectors (e.g. face detectors) can offer varying degrees of discrimination, robustness and persistence, depending on their misses and false positives.

In the following sections such cues are detailed, focusing on:

- Analysing the different measurement cues, considering the different information each of them can bring to the tracker and their suitability for tracker initialisation.
- Demonstrating the different measurement cues with example code.
- Constructing likelihood models $p\left(\mathbf{y}^{(c)}|\mathbf{x}\right)$ for all cues c, where $\mathbf{y}^{(c)}$ is the measurement for cue c and $\mathbf{x}$ is the state vector.
- Paving the way for efficient proposal distribution design using measurements in section 4.11.3.

The first two goals are independent of the tracking algorithm, the third refers to stochastic tracking, while the fourth is related to particle filter tracking only.

Measurements for visual tracking are made using image and video signal processing techniques. They all attempt to abstract just enough information from the target to allow tracking without ambiguity from changes in the target appearance. Such changes occur due to both changing imaging conditions and flexing of the target. At the one end, all (or almost all) spatial information is abstracted using colour: particular coloured regions can be tracked in images by matching pixels or entire regions to a colour histogram (section 4.5). At the other end, all spatial information is retained when outlines resembling the target shape are matched to image regions by taking advantage of the contrast change between targets and the background (section 4.6). In setups where cameras are static, moving targets can be discriminated from the background based on their motion (section 4.7). Finally, some targets such as faces of people are quite distinct. They can be detected given favourable viewing conditions (i.e. nearly frontal pose, neutral expression, uniform illumination and adequate resolution), yielding the position of human heads (section 4.8).

Due to the different weaknesses of all visual cues, robust visual measurements involve multiple complementary cues. For example the persistence and robustness of colour can be complemented with the discriminative ability of motion and outline matching. The ways to involve multiple cues in multimodal trackers are explored in Chapter 5.

4.5   Colour matching

Colour is an important property of visual targets. The colours of a target are very robust to changes in viewing conditions such as illumination, pose and flexing of the target. They are also very persistent: as long as the target is visible, the colours are there to be measured. Only occlusions can hide them. On the other hand, the colours of a target may not be unique and can also appear in the background. Hence colour is neither discriminative, nor does it have good detection properties. Ways to use colour for detection validation are described in section 4.8.2.

There are two options regarding the use of colour:

- Discard all the target's spatial information by using colour-only matching. This is the option for targets with rather uniform colours. Skin patches are typical examples.
- Keep part of the spatial information, by using colour features enhanced with other information. This is meaningful for targets that have different colours in different positions. The upper human body is a typical example, with the following zones from top to bottom: hair, skin-like face and upper torso clothing.

In the following subsections colour matching for both cases is analysed.

4.5.1 *Colour without spatial information*

Measurement of the colour cue, as for all other cues, is done in the image region $R_\mathbf{x}$ dictated by the state $\mathbf{x}$ via the mapping $\mathbf{x} \to R_\mathbf{x}$. In its simplest form, colour measurement involves the comparison of the colour values of the pixels in $R_\mathbf{x}$ against some limits. For example, human skin has stronger red component than green or blue. The existence of multiple colours in the target and colour variations due to lighting leads to complicated limits of the colour components. In such cases it is simpler to evaluate the colours of $R_\mathbf{x}$ against some reference colour histogram, as first introduced in [Swain and Ballard (1991)].

One approach is to build independent red, green and blue colour component reference histograms [Pérez *et al.* (2004)]. A more efficient approach is to take into account the correlation between colour components by building a joint histogram of the three colour components [Pérez *et al.* (2002)]. To do so efficiently, the red (R), green (G) and blue (B) colour components are first quantised to N_h levels and then are combined into a one-dimensional colour quantity $c(R, G, B)$, defined as:

$$c(R, G, B) \equiv \left\lfloor R \frac{N_h}{256} \right\rfloor + N_h \left\lfloor G \frac{N_h}{256} \right\rfloor + N_h^2 \left\lfloor B \frac{N_h}{256} \right\rfloor \qquad (4.8)$$

where $\lfloor a \rfloor$ denotes the largest integer smaller than or equal to a. The histogram of $c(R, G, B)$ is then calculated. Since $\{R, G, B\} \in [0, 255]$,

$c(R, G, B)$ is an integer in the range $[0, \ldots, N_h^3 - 1]$, yielding N_h^3 histogram bins.

Example 4.3. The following MATLAB function extracts a histogram h from a rectangular image region Rx using Nh quantisation levels per colour component.

```
function h=histogram(Rx,Nh)
Rr=double(reshape(Rx,numel(Rx(:,:,1)),3));
c=floor(Rr(:,1)*Nh/256)+Nh*floor(Rr(:,2)*Nh/256)+Nh2*floor(Rr(:,3)*Nh/256);
h=hist(c,0:Nh^3-1);
h=h/sum(h);
```

∎

Matching of the image region $R_\mathbf{x}$ against the reference histogram $\mathbf{h}_\text{ref}$ can be done for every pixel in $R_\mathbf{x}$. Given the pixel's colour quantity $c(R, G, B)$, the evaluation of the reference histogram at $c(R, G, B)$ gives the probability that the pixel has a colour belonging to the target. A probability threshold T_h can be used to accept the pixel as belonging to the target if

$$\mathbf{h}_\text{ref}\left(c(R, G, B)\right) > T_h \tag{4.9}$$

The pixel-based distance $D(R_\mathbf{x}, \mathbf{h}_\text{ref})$ of $R_\mathbf{x}$ from the target colour histogram is given by the ratio of the number of pixels not fulfiling (4.9), over the total number of pixels in $R_\mathbf{x}$. $D(R_\mathbf{x}, \mathbf{h}_\text{ref})$ is a proper metric, since it ranges between 0 for all pixels in $R_\mathbf{x}$ fulfiling (4.9) to 1 for no pixels in $R_\mathbf{x}$ fulfilling (4.9).

The problem with the use of pixel-level matching is the need to define the probability threshold T_h. Such thresholds are empirically defined, and their optimum values are application-dependent. A neater approach is to use region-level matching, by constructing a target histogram $\mathbf{h}_\mathbf{x}$ from the pixels in $R_\mathbf{x}$ and comparing it to the reference histogram using the Bhattacharyya distance:

$$D(\mathbf{h}_\mathbf{x}, \mathbf{h}_\text{ref}) = 1 - \sum_{n=0}^{N_h^3-1} \sqrt{\mathbf{h}_\mathbf{x}(n)\mathbf{h}_\text{ref}(n)} \tag{4.10}$$

where $\mathbf{h}_\mathbf{x}(n)$ and $\mathbf{h}_\text{ref}(n)$ are the n-th bins of the target and reference histograms. Similar to the pixel-level matching distance, the region-level matching distance $D(\mathbf{h}_\mathbf{x}, \mathbf{h}_\text{ref})$ is also a proper metric, since it ranges between 0 for the two histograms being the same, to 1 for the two histograms having no common non-zero bin.

No matter the colour matching approach, pixel- or region-level, the likelihood for the colour measurement $\mathbf{y}^{(\text{colour})}$ is defined as :

$$p\left(\mathbf{y}^{(\text{colour})}|\mathbf{x}\right) \propto \exp\left(-D/2\sigma_{\text{colour}}^2\right) \qquad (4.11)$$

where D is either the pixel-based $D(R_\mathbf{x}, \mathbf{h}_{\text{ref}})$ or the region-based Bhattacharyya $D(\mathbf{h}_\mathbf{x}, \mathbf{h}_{\text{ref}})$ and σ_{colour}^2 is the colour likelihood variance. The exponential distribution of the colour matching distance is justified experimentally [Pérez *et al.* (2004)].

Example 4.4. Evaluation of the colour likelihood $p\left(\mathbf{y}^{(\text{colour})}|\mathbf{x}\right)$ in (4.11).

We use the image `001.jpg` depicting a meeting between three people (shown in Fig. 4.5), and the human skin colour histogram in the file `skin.hist`. This is the histogram described in [Jones and Rehg (2002)], with $N_h = 32$. To load and normalise the reference histogram `href` in MATLAB, use:

```
href=load('skin.hist');
href=href/sum(href);
```

Assume that the state is just the horizontal and vertical image coordinates, $\mathbf{x} = [x, y]^T$ and that the mapping $\mathbf{x} \to R_\mathbf{x}$ maps the state into a rectangle with top-left corner $\mathbf{x}$, fixed width W and height $H = 4W/3$. Note that this is the face aspect ratio, hence the colour likelihood should be a good choice for colour-based face tracking.

Given the image `I`, the width `W`, the height `H`, the colour likelihood standard deviation `s2c` and the horizontal and vertical step for scanning the image `st`, the colour likelihood `pyx` is obtained on a subsampled grid using the following MATLAB code:

```
pyx=zeros(floor((size(I,1)-H)/st)+1,floor((size(I,2)-W)/st)+1);
for m=1:st:size(I,1)-H+1
    for n=1:st:size(I,2)-W+1
        Rx=I(m:m+H-1,n:n+W-1,:);
        hx=histogram(Rx,32);
        D=1-sum(sqrt(hx.*href));
        pyx((m-1)/st+1,(n-1)/st+1)=exp(-D/(2*s2c));
    end
end
```

where the function `histogram()` has been defined in example 4.3. Note that the number of quantisation levels per colour component N_h needs to be the same (i.e. 32) in the reference and the target histograms. colour information is matched at the region-level, using the Bhattacharyya distance of (4.10).

Colour likelihood assumes exponential distribution of the distance, as in (4.11).

The image and the resulting colour likelihood are shown in Fig. 4.5. The colour likelihood is evaluated with a step of 0.625% of the image width. The following are to be noted:

- The face patches give large likelihoods, which is good for face tracking. The likelihood at the well lit face closer to the camera is larger than the likelihoods of the shadowed faces.
- All other skin patches (hands, arms) give large likelihoods, their value depending on the size of the skin patch relative to the size of $R_{\mathbf{x}}$. If they are significantly smaller than the target search rectangle, then colours that are not skin-like can cause large matching distances.
- There are many patches with colours similar to human skin (tabletops, floor, walls), that cause large likelihoods, even larger than those of some of the skin or face patches.

These observations demonstrate that even though colour is persistent (the faces always cause large colour likelihoods), it is not a suitable cue for tracker initialisation (it can initialise a target on a table top in this example), and it can easily result to ambiguity (the track on the face at the centre might be lost as the tracker can stick to the larger likelihood of the table top on its left). ∎

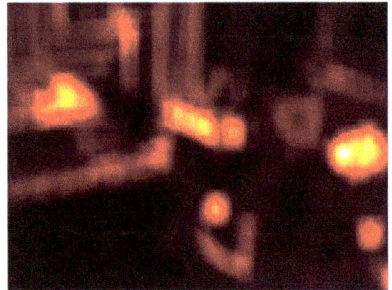

Fig. 4.5   Image and associated colour likelihood with a generic human skin colour histogram as reference.

4.5.2 *Colour with spatial information*

In example 4.4, the assumption is that the target mainly comprises human skin. This is not entirely true for faces, as the eye region, the possible existence of glasses, facial hair and make-up can lead to regions with grossly different colours. Things are certainly so if the target is not the face but the entire head. In such cases, all the colours are described by the histogram but any information on the spatial arrangement of the colours inside the target is lost. The remainder of this section describes ways to keep this spatial information to some degree.

Assume that the target contains n_r distinct colour regions, each of area $A^{(i)}$, $i = 1, \ldots, n_r$. Also assume that the spatial arrangement of these regions is known, either fixed (non-flexing targets) or flexible (modelled by some state variables for flexing targets). Then n_r reference histograms $\mathbf{h}_{\text{ref}}^{(i)}$ are trained and are compared to the target histograms $\mathbf{h}_{\mathbf{x}}^{(i)}$. An overall distance $D_{\text{tar,ref}}$ for the multi-region colour likelihood can be defined as the weighted average of the distances $D^{(i)}$ of the per-region reference and target histograms:

$$D_{\text{tar,ref}} = \sum_{i=1}^{n_r} \frac{A^{(i)}}{\sum_{k=1}^{n_r} A^{(k)}} D^{(i)} \tag{4.12}$$

The weights are chosen in proportion to the area of each region. This effectively assigns an importance to how well each of them is matched that is proportional to their size. As for the per-region distances $D^{(i)}$, they can either be the pixel-based $D\left(R_{\mathbf{x}}^{(i)}, \mathbf{h}_{\text{ref}}^{(i)}\right)$ or the Bhattacharrya $D\left(\mathbf{h}_{\mathbf{x}}^{(i)}, \mathbf{h}_{\text{ref}}^{(i)}\right)$. In the multi-region case, the colour likelihood is again given by (4.11) only the distance D is replaced by $D_{\text{tar,ref}}$.

Example 4.5. Evaluation of the colour likelihood for a target with multiple colour regions.

We use the image 001.jpg (shown in Fig. 4.5) to train a colour histogram suitable for tracking human heads. There are three heads visible there. Using their pixels, we train three different histograms:

- A face histogram using the face pixels.
- A single-region head histogram using the head region, defined as the face rectangle with an extra hair region of one-sixth of the height of the face above it.
- A dual-region head histogram using the face and hair regions independently.

The two regions making up each of the three heads used to train the histograms are shown in Fig. 4.6.

The state is the same used in example 4.4, with constant width W being the mean of the widths of the training faces, and height $H = 4W/3$ for faces, $H = 4 * 7W/(3 * 6)$ for heads and $H = 4W/(3 * 6)$ for hair. The colour likelihood is evaluated on the same subsampled grid, only using a different image than for training, image 002.jpg, depicting the same scene but from a different angle (shown in Fig. 4.6). If the reference face and hair histograms are stored in **href2** and **href1** respectively, the likelihood for the dual region target representation is calculated as follows:

```
pyx=zeros(floor((size(I,1)-H)/st)+1,floor((size(I,2)-W)/st)+1);
for m=1:st:size(I,1)-H+1
    for n=1:st:size(I,2)-W+1
        Rx1=I(m+round(H/7):m+H-1,n:n+W-1,:);
        Rx2=I(m:m+round(H/7)-1,n:n+W-1,:);
        hx1=histogram(Rx1,Nh);
        hx2=histogram(Rx2,Nh);
        D1=1-sum(sqrt(hx1.*href1));
        D2=1-sum(sqrt(hx2.*href2));
        D=(6*D1+D2)/7;
        pyx((m-1)/st+1,(n-1)/st+1)=exp(-D/(2*s2c));
    end
end
```

Fig. 4.7 depicts the frame itself and the colour likelihoods obtained by using the histograms trained from the faces, the heads and the faces and hair as separate regions. In order to validate the success of each approach, there are two factors to consider:

- Colour likelihood of objects not belonging to the class of tracked objects should be significantly smaller than that of those belonging to the class of tracked objects. This reduces the effect of clutter from the background and facilitates the use of the likelihood for track initialisation.
- Colour likelihood of objects belonging to the class of tracked objects should be similar. This facilitates the use of the likelihood for track initialisation.

To evaluate the selection for the reference histogram, we consider the two objects belonging to the class of tracked objects, i.e. the two heads, and two objects that have skin-like colour, the hands of the person on the right and the table top over his face. The two faces should have comparable likelihoods, much larger than those of the hands and table top. Table 4.1 shows the likelihoods for the left face, the hands and the table top, normalised by that of the right face, for all three trained histograms. The following can be observed for the three histograms:

- Face colour histogram. In this case the hands have greater likelihood than that of the right face, while the table top is much lower. The left face has the dominant likelihood. Using this single region histogram for track initialisation is not possible. Even for tracking, there is significant ambiguity.
- Head colour histogram. In this case the hands have smaller but comparable likelihood than that of the right head, while the table top likelihood is somewhat increased. The left head is still dominant, albeit with smaller distance. Using this single region histogram for track initialisation is again not possible, but for tracking, the ambiguity is reduced.
- Face and hair colour histograms. Now all background objects have significantly smaller likelihood than the heads. The left head has comparable likelihood to the right one. Using this dual region histogram is suitable both for track initialisation and for tracking.

■

Fig. 4.6   Training colour histograms using three heads, split into face and hair regions.

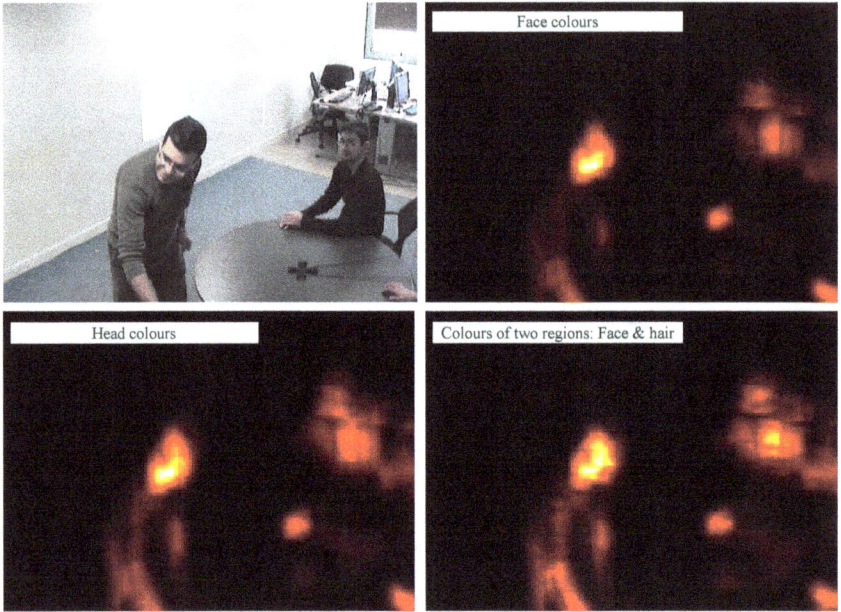

Fig. 4.7   Image and associated three subsampled colour likelihoods with reference histograms trained from face colours, head colours or dual region face and hair colours respectively.

Table 4.1   Maximum colour likelihood of different objects (in rows) for reference histograms based on single or multiple regions and only colour or colour jointly with distance (in columns). The three first histograms refer to example 4.5, and the fourth to example 4.6.

Objects	One region: Face	One region: Head	Two regions: Face+hair	Joint: Colour+distance
Right head	1	1	1	1
Left head	2.32	1.93	1.27	1.18
Hands	1.21	0.989	0.655	0.568
Table top or floor	0.612	0.723	0.495	0.492

Note that there are other ways to include target structure in the histograms, utilising higher dimensional features. One approach is to augment the colour vector with position (x, y) or distance $r = \sqrt{x^2 + y^2}$ relative to an anchor point, leading to five or four dimensional features respectively. The extra dimensions are quantised in a way similar to the colour triplet, and all of them are combined into a one-dimensional quantity as in (4.8). If the n-dimensional feature vector used is $\mathbf{v} = [v_1, \dots, v_n]$, with $v_i \in \left[0, v_i^{(\max)}\right)$, and N_h is the number of quantisation levels per feature, then the quantity c becomes:

$$c(\mathbf{v}) = \sum_{i=1}^{n} N_h^{i-1} \left\lfloor v_i \frac{N_h}{v_i^{(\max)}} \right\rfloor \tag{4.13}$$

The drawback is that the number of histogram bins increases exponentially with the number n of the features used, since it is N_h^n. Hence the processing load of evaluating the Bhattacharrya distance increases dramatically with n.

Example 4.6. Evaluation of the joint colour and distance from centre likelihood for a target with multiple colour regions.

The following MATLAB function extracts a joint colour-space histogram h from a rectangular image region Rx using Nh quantisation levels per feature. The features used are three for the colour components and one for the distance of the pixel from the centre of Rx.

```
function h=histogramCS(Rx,Nh)
Rr=double(reshape(Rx,numel(Rx(:,:,1)),3));
[y x]=ind2sub(size(Rx(:,:,1)),1:numel(Rx(:,:,1)));
r=sqrt((x-mean(x)).^2+(y-mean(y)).^2)';
Rr=[Rr r];
v=[256 256 256 sqrt(size(Rx,1)^2+size(Rx,2)^2)/2];
c=zeros(size(Rr,1),1);
for i=1:size(Rr,2)
    c=c+Nh^(i-1)*floor(Rr(:,i)*Nh/v(i));
end
h=hist(c,0:Nh^size(Rr,2)-1);
h=h/sum(h);
```

Note that while the possible range for the colour components is the full scale, no matter whether such colours are present, the full scale for the distance from centre is adjusted to the size of Rx. This is achieved by setting $v_4^{(\max)} = \sqrt{a^2 + b^2}/2$, where a and b are the width and height of Rx. The effect of the adjustment is to have colour-space histograms that are independent of the size of the region.

The resulting histogram is used to evaluate the joint colour and distance likelihood for the image 002.jpg (shown in Fig. 4.6). The likelihood is shown in Fig. 4.8, and the relative likelihoods of the different objects are given in the last column of Table 4.1. Obviously the joint colour and distance histogram performs better than the colour-only ones. ∎

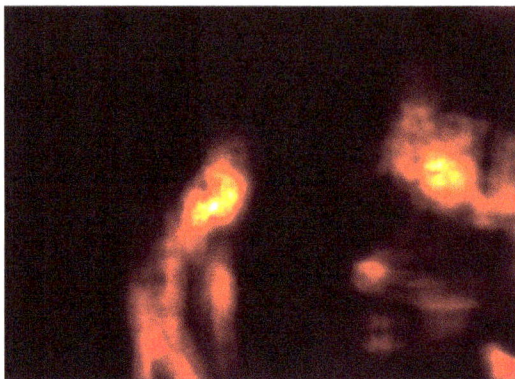

Fig. 4.8   Joint colour and distance likelihood for the image of Fig. 4.7 and the histogram trained from the heads of Fig. 4.6.

While the use of a joint colour and space histogram improves both the tracking and detection capabilities over colour-only histograms, the processing load increases dramatically. For both examples 4.5 and 4.6, the time needed for the evaluation of the likelihood normalised by that of the single-region (face) colour-only histogram is 1.06 for the single-region (head) colour-only one, 1.80 for the dual-region (hair and face) colour-only one and 19.5 for the single-region (head) joint colour-space one.

Other approaches to include target structure in the histograms are the use of histograms of the ratios of neighbouring pixels [Funt and Finlayson (1995)] and of colour co-occurrence histograms [Chang and Krumm (1999)].

4.6   Outline matching

Subsection 4.5.2 attempted to introduce some structure information into the colour-based description of the target. The use of target outlines on the other hand abstracts away all colour information, keeping all structure information.

If the target has a distinctive shape, then it is advantageous to learn a model of target outlines from example images. Once this model captures all the possible shapes of the target, it can be very powerful. On the other hand, there are two major difficulties with using target outlines:

- Learning the outline model. This can be very difficult as the shape of the target can change due to both viewpoint changes and target flexing. Viewpoint changes for (almost) rigid targets can be accounted for by modelling the outline in 3D: the target's 3D position and orientation are included in the state. The outline is then projected onto the camera view, using the camera's calibration parameters (subsection 4.6.2). Target flexing can be partly handled by learning complicated shape models using active contours [Blake and Isard (1998)].

- Background clutter. Target outlines are matched utilising gradient information as evidence. Target contrast from the background can produce such evidence, but there is no guarantee that even stronger false evidence does not exist in the background near the target or within the target itself. The clutter can be partly accounted for using edge orientation (subsection 4.6.1) or motion (subsection 4.7.1) information.

Outline matching and its various options are detailed in this section using the upper human torso as an example. The only sources of flexing of this part of the human body are the neck and the shoulders. The head stays mostly upright in many applications, as it only occasionally bends. Head rotation does not change the outline significantly. The shoulders are raised in some rare gestures, again not changing the upper torso outline significantly. A suitable outline model for the human upper torso, including the shoulders, neck and head, comprising 24 line segments symmetrically located around the y-axis is shown in Fig. 4.9, and is fit onto a frontal human image. To achieve this, the model is shifted and scaled accordingly. The horizontal (x) and vertical (y) shifts and the scale (a) are the three variables of the state vector $\mathbf{x}$. Assuming that the outline model comprises K points, $\mathbf{x}_m^{(i)} = \left[x_m^{(i)}, y_m^{(i)} \right]^T$, $i = 1, \ldots, K$, the mapping of the state into the K image pixels $\mathbf{x}_p^{(i)}$ given the model, $\mathbf{x} \rightarrow \mathbf{x}_p^{(i)} | \mathbf{x}_m^{(i)}$ is:

$$\mathbf{x}_p^{(i)} = a \cdot \left[x_m^{(i)}, y_m^{(i)} \right]^T + [x, y]^T \tag{4.14}$$

Line segments perpendicular to the outline are utilised to fit the outline onto the image of a target. They are centred around key-points of the outline. Evidence to support the assumed position of the outline is gathered along these segments [Isard and Blake (1998)]. The evidence can either be a steep gradient of the intensity along the segments, or pixels of edges crossing the segments. These two options are shown in Fig. 4.10.

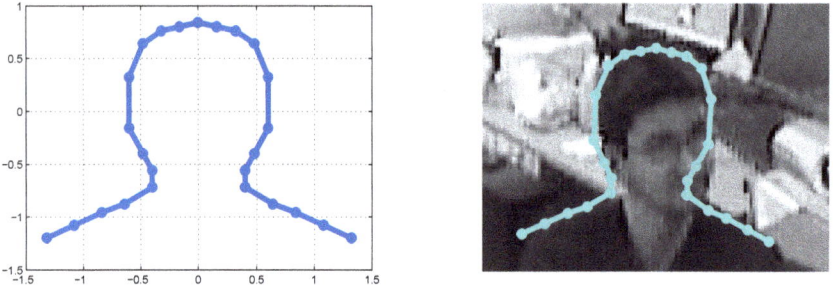

Fig. 4.9 Outline of upper human torso and its fit via the state on the image of a frontal torso with somewhat rotated head. The outline is tightly fit around the head and shoulders, but not the neck. There, the rotation of the head causes the outline to be far from the target on the left hand side, and inside the target on the right hand side.

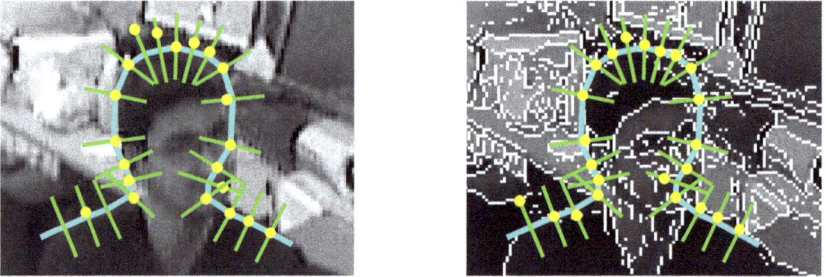

Fig. 4.10 Fitting the outline on an image by searching for supporting evidence along lines perpendicular to the outline. (Left) 1D approach: searching for substantial grey-level change along the perpendicular lines segments. (Right) 2D approach: searching for edge pixels along the segments.

When such evidence is found close to the middle of the perpendicular line segments, then the fit of the outline on the candidate image (defined by the state $\mathbf{x}$) is considered better. To quantify the quality of the fit, assume N segments, and distance d_i pixels, $i = 1, \ldots, N$, of the nearest evidence from the centre of segment i. The distance $D_{\text{outline}}(\mathbf{x})$ of the outline mapped on the image at state $\mathbf{x}$ from all supporting evidence is defined as the mean of the d_i. To make this distance a proper metric, the maximum value has to be unity. To make it so, all the d_i have to be normalised by the maximum distance from the centre of the perpendicular segments, which is their half-length L. Hence the distance is defined as:

$$D_{\text{outline}}(\mathbf{x}) = \frac{1}{N\dot{L}} \sum_{i=1}^{N} d_i \qquad (4.15)$$

The collection of matching evidence and the calculation of the distances d_i are detailed in the example 4.7.

Example 4.7. The following MATLAB script calculates the matching distances d_i (denoted Di) for all the segments of the outline. It requires the following input:

- xp: the coordinates of the corner points of the outline as it is projected onto the image position under evaluation.
- L: the half-length of the line segments that are perpendicular to the outline and along which matching information is collected. It is usually obtained as a fraction (e.g. 10%) of the diagonal of the bounding box containing xp.
- Lbb: a rectangle containing xp, obtained by extending the bounding box of xp by a multiple of L (e.g. 120%) across every direction.
- I: the pixels within Lbb. These have either the grey-scale values or the edges of the original image bounded by Lbb.

The script spans the outline corner points, omitting the first and the last. In the loop, the following are performed:

- Calculation of the line segment that is centred around the p-th corner point of the outline and is perpendicular to the line connecting the p-1 and p+1 corner points.
- Collection of evidence along the line segments. This is done by first cropping the relevant portion of I and then finding the pixel-value cross-sections (pr) along the p-th line segment. Finally, those of the pr points for which there is matching evidence are identified, as detailed later in the example. The indices of these points are denoted ind.
- Calculation of the p-1 element of the Di as the smallest distance of all the ind points of the pr from the p-th corner point. If there are no pr points with adequate evidence, the distance is set to the half-length of the segments.

```
% For every point of the outline
Di=zeros(1,size(xp,2)-2);
for p=2:size(xp,2)-1
```

```
% Find prependicular segment endpoints (x1,y1)
% and bounding box (xlow,xhigh,ylow,yhigh)
xo=xp(1,p-1:p+1);
yo=xp(2,p-1:p+1);
phiM=atan((yo(1)-yo(3))/(xo(1)-xo(3)));
a=tan(pi/2+phiM);
if a<100
    b=yo(2)-a*xo(2);
    x1=xo(2)+L/sqrt(a^2+1)*[-1 1];
    y1=a*x1+b;
else
    y1=[yo(2)-L yo(2)+L];
    x1=[xo(2) xo(2)];
end
y1(y1<1)=1;
y1(y1>size(I,1))=size(I,1);
x1(x1<1)=1;
x1(x1>size(I,2))=size(I,2);
xlow=max([1 floor(min(x1))-1]);
xhigh=min([size(I,2) ceil(max(x1))+1]);
ylow=max([1 floor(min(y1))-1]);
yhigh=min([size(I,1) ceil(max(y1))+1]);
% Find evidence along line segments
Ic=I(ylow-Lbb(2,1)+1:yhigh-Lbb(2,1)+1,...
    xlow-Lbb(1,1)+1:xhigh-Lbb(1,1)+1);
[cx cy pr]=improfile(Ic,x1-xlow+1,y1-ylow+1,31,'bilinear');
% For which of the pr is there matching evidence?
ind=matchGrey(pr);
% Minimum distance from outline
if ~isempty(ind)
    Di(p-1)=min(abs(ind-(size(pr,1)-1)/2-1));
else
    Di(p-1)=round(size(pr,1)/2+1);
end
end
```

In order to find which of the values **pr** sampled along each of the perpendicular segments actually indicate a match, function `matchGrey()` is used for finding large grey-scale changes:

```
function ind=matchGrey(pr)
pr=abs(filter([-1 0 1],1,[pr;pr(end,:);pr(end,:)]));pr=pr(3:end,:);
ind=find(max(pr>20,[],2));
```

In this case I is the grey-scale image segment in **Lbb**, and **pr** is obtained by 1D processing along the line segment.

The matching function is changed to `matchEdge()` in order to find the presence of adequate edge information:

```
function ind=matchEdge(pr)
ind=find(pr>.1);
```

In this case I is the edge image obtained by the 2D image processing necessary for edge detection [Gonzalez and Woods (2007)]. ∎

The likelihood function for outline matching is defined as an exponentially decaying function of $D_{\mathrm{outline}}(\mathbf{x})$, in the same manner as for colour:

$$p\left(\mathbf{y}^{(\mathrm{outline})}|\mathbf{x}\right) \propto \exp\left(-D_{\mathrm{outline}}(\mathbf{x})/2\sigma_{\mathrm{outline}}^2\right) \qquad (4.16)$$

An example of the likelihood function for edge pixels serving as evidence is shown on the top-right image of Fig. 4.11.

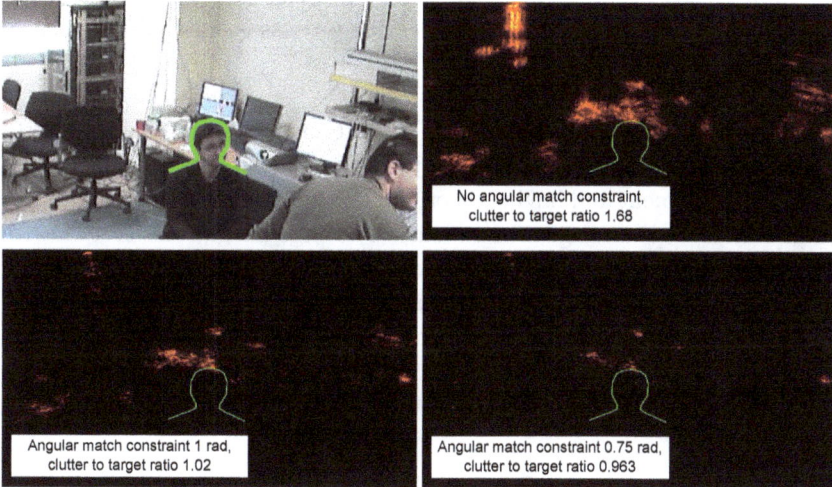

Fig. 4.11   Image with best fit of upper torso outline overlaid (top-left) and associated three likelihood functions. Top-right: without any evidence vs. outline orientation match constraint. Bottom-left: 1 rad relaxed orientation match constraint. Bottom-right: 0.75 rad stricter orientation match constraint.

4.6.1   *Clutter attenuation by edge orientation*

It is evident from the likelihood depicted at the top-right of Fig. 4.11 that the upper torso outline can be fitted around the head and shoulders of the person yielding large likelihood, but the image clutter results into too many edges, some of which can also be matched to the outline, with even larger likelihoods.

When the collected evidence is edge information, then there is a straight-forward solution to reduce the matching false positives: the matching criterion is refined by asking for the edges that contribute to the evidence to have similar orientation as the outline at the given point. This way some of the clutter edges will be disregarded, as their orientations will not match that of the outline.

Matching of the orientations is tested by comparing the angle ϕ_i of the outline at the particular point i (obtained as the slope of the line connecting the previous and the next point) with the angle of the candidate evidence. The evidence angle $\phi^{(e)}$ is estimated as follows: a 5×5 neighbourhood is centred around the cross-section of the edge and the perpendicular line segment. Inside the neighbourhood the connected component of edge pixels containing the cross-section is identified, and the best linear fit of those pixels is calculated, yielding the angle. The process is detailed in Fig. 4.12 and example 4.8.

Example 4.8. The following MATLAB function can replace `matchGrey()` in the code of example 4.7 to find the presence of edges crossing the perpendicular line segment and having an orientation not different from that of the outline by more than `angleThr` radians. The edge image cross-section with the line segment is `pr`. The outline angle is `phiM` (this is already calculated to find the perpendicular line segment). Finally, the edge image cropped around the perpendicular line segment is `Ic`.

```
function ind=matchEdgeAngle(pr,Ic,angleThr,phiM)
% Work with detected edges
ind=find(pr>.1);
% Find match of edge and outline orientations
[r c]=find(Ic);
angleMatch=zeros(1,length(ind));
for i=1:length(ind)
    [m im]=min((r-cy(ind(i))).^2+(c-cx(ind(i))).^2);
    Icc=Ic(max([1 r(im)-2]):min([size(Ic,1) r(im)+2]),...
            max([1 c(im)-2]):min([size(Ic,2) c(im)+2]));
    IccL=bwlabel(Icc);
    Icc=IccL==IccL(13);
    [rm cm]=find(Icc);
    pol=polyfit(cm,rm,1);
    angleMatch(i)=abs(atan(pol(1))-phiM);
end
ind
ind=ind(angleMatch<angleThr);
```

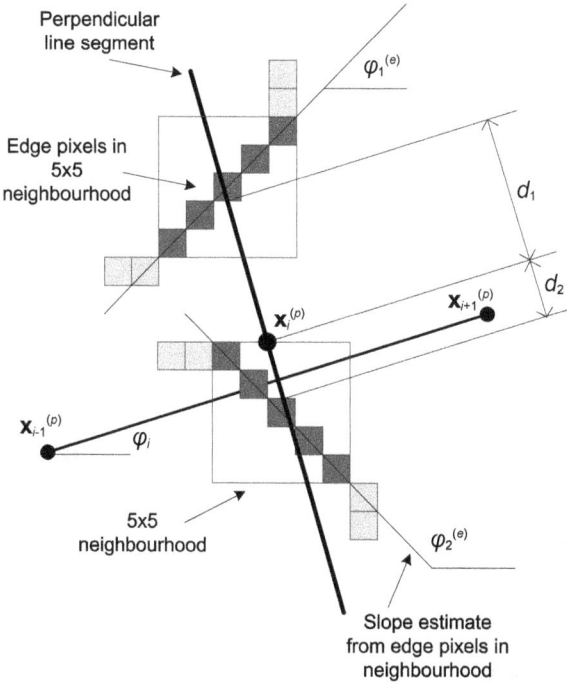

Fig. 4.12 Matching the outline against edge evidence using an orientation constraint. Although d_2 is smaller than d_1, the matching is done against the upper edge, if $\phi_2^{(e)} - \phi_i$ is larger than the maximum allowed orientation difference and $\phi_1^{(e)} - \phi_i$ is smaller than it.

The success of bounding the maximum allowed orientation difference on attenuating the clutter can be seen qualitatively in the likelihoods depicted at the bottom row of Fig. 4.11. It can also be quantified using the percentage of pixels with large clutter to target ratio. As seen in Fig. 4.13, moderate bounds on the maximum allowed orientation difference reduce this percentage drastically. Bounds stricter than about 40° start having the opposite effect, as now most of the candidate edge matches are discarded,leading to overall increased reported outline distances.

4.6.2 *View changes handling by modelling in 3D*

The outline of a target changes a lot as the viewpoint changes. All the upper torso outline matching algorithms presented thus far assume that the target is (approximately) facing the camera. When this is not the case,

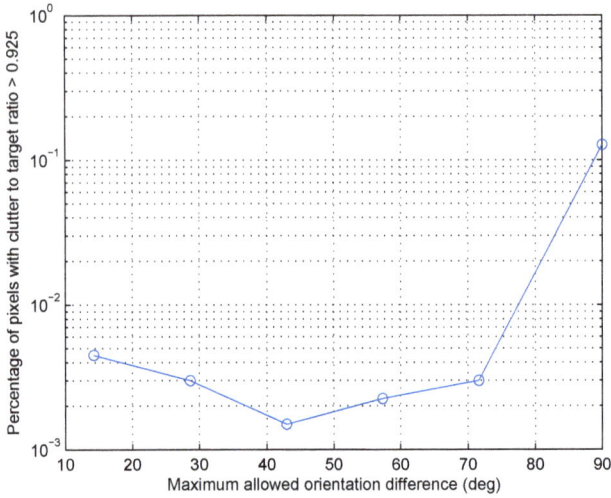

Fig. 4.13   Percentage of pixels with clutter to target ratio larger than 0.925, as a function of the maximum allowed orientation difference.

the outline of Fig. 4.9 needs to be projectively transformed into a shape that matches the upper human torso under the given rotation.

This can be readily accomplished by using an outline state in the 3D space, as shown in Fig. 4.14. Compared to the state on the image plane, the one in the 3D space utilises three shift components (x and y along the floor-plan and h for height) and an orientation component ϕ, which is the angle of rotation along the z axis, or equivalently the vector $\mathbf{f}$ that is perpendicular to the outline plane w.r.t. the x axis. The scale component remains as is.

The matching likelihood is estimated as follows:

- The outline is transformed according to the state into the 3D world coordinate system.
- The 3D outline is projected onto all available calibrated camera views.
- The likelihood function for outline matching per camera view $p\left(\mathbf{y}^{(\text{outline,camera})}|\mathbf{x}\right)$ is calculated as discussed earlier in this section.
- The overall outline matching likelihood is calculated from the per camera view likelihoods.

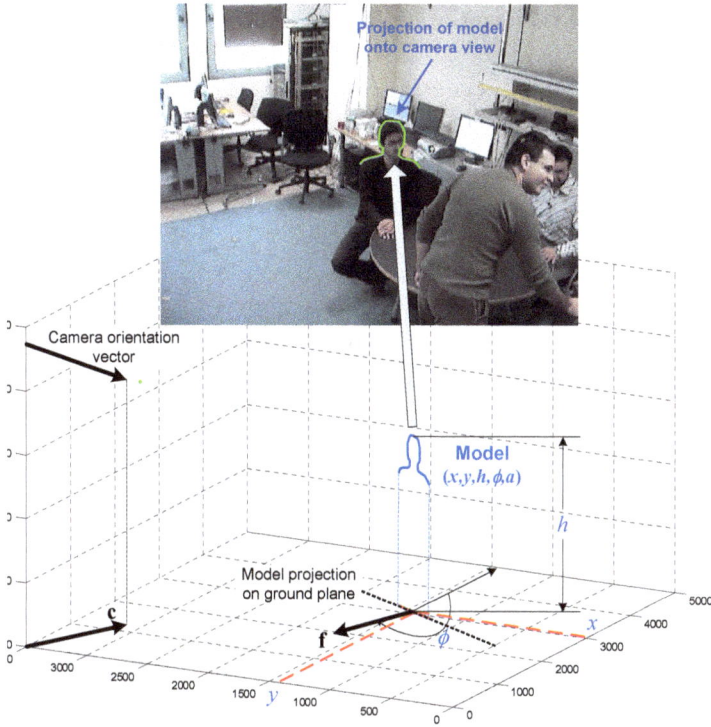

Fig. 4.14   Definition of the state of an upper torso outline in the 3D space and mapping onto an image.

The outline of Fig. 4.9, defined at the y-z plane is given by the K points whose homogeneous 3D coordinates are given by the following four-element vectors:

$$\mathbf{x}_m^{(i)} = \left[0, y_m^{(i)}, z_m^{(i)}, 1\right]^T \qquad (4.17)$$

where $i = 1, \ldots, K$, the coordinate along the x axis is zero and the 3D coordinates are turned into the equivalent homogeneous ones by appending by a unity fourth element. The homogeneous coordinates facilitate the use of 4×4 3D transformation matrices for the transformation of the model $\mathbf{x}_m^{(i)}$ from the y-z plane into the 3D space $(\mathbf{x}_{3D}^{(i)})$ [Molhave (2000)] with the appropriate orientation, scale and translation, according to the state $\mathbf{x} = [x, y, h, \phi, a]^T$. Three 3D transformation matrices are needed: one for the rotation along the z axis by ϕ ($\mathbf{R}_z$), another for the scaling by a ($\mathbf{S}$)

and a final one for the translation by x, y and h along the x, y and z axes respectively ($\mathbf{T}$), where:

$$\mathbf{R}_z = \begin{bmatrix} \cos\phi & \sin\phi & 0 & 0 \\ -\sin\phi & a\cos\phi & 0 & 0 \\ 0 & 0 & 1 & 0 \\ 0 & 0 & 0 & 1 \end{bmatrix} \tag{4.18}$$

$$\mathbf{S} = \begin{bmatrix} a & 0 & 0 & 0 \\ 0 & a & 0 & 0 \\ 0 & 0 & a & 0 \\ 0 & 0 & 0 & 1 \end{bmatrix} \tag{4.19}$$

and

$$\mathbf{T} = \begin{bmatrix} 1 & 0 & 0 & 0 \\ 0 & 1 & 0 & 0 \\ 0 & 0 & 1 & 0 \\ x & y & h & 1 \end{bmatrix} \tag{4.20}$$

Then the transformation is given by:

$$\mathbf{x}_{3D}^{(i)} = \mathbf{T}^T \left(\mathbf{S}^T \left(\mathbf{R}_z^T \mathbf{x}_m^{(i)} \right) \right) = \left((\mathbf{R}_z \mathbf{S}) \, \mathbf{T} \right)^T \mathbf{x}_m^{(i)} = \mathbf{W}^T \mathbf{x}_m^{(i)} \tag{4.21}$$

where:

$$\mathbf{W} \equiv (\mathbf{R}_z \mathbf{S}) \, \mathbf{T} = \begin{bmatrix} a\cos\phi & a\sin\phi & 0 & 0 \\ -a\sin\phi & a\cos\phi & 0 & 0 \\ 0 & 0 & a & 0 \\ x & y & h & 1 \end{bmatrix} \tag{4.22}$$

Example 4.9. Project the upper torso outline $\mathbf{x}_m^{(i)}$ from the y-z plane into the 3D space yielding $\mathbf{x}_{3D}^{(i)}$ using the following four states concatenated into matrix $\mathbf{X}$:

$$\mathbf{X} = \begin{bmatrix} 640 & 3500 & 1930 & 3110 \\ 1410 & 2580 & 3700 & 3680 \\ 1670 & 1720 & 1600 & 1680 \\ \frac{\pi}{6} & \frac{\pi}{3} & \frac{2\pi}{3} & \frac{\pi}{3} \\ 165 & 165 & 165 & 160 \end{bmatrix} \tag{4.23}$$

The following MATLAB code generates $\mathbf{W}$ from every column of $\mathbf{X}$ (state) according to (4.22) and then calculates the four K-point 3D outlines in homogeneous coordinates $\mathbf{x}_{3D}^{(i)}$ (stored in the $4 \times K \times 4$ array x3D) by projecting $\mathbf{x}_m^{(i)}$ (stored in the $4 \times K$ array xm) as in (4.21):

```
for k=1:size(X,2)
    W=[X(5,k)*cos(X(4,k))  X(5,k)*sin(X(4,k)) 0      0;...
       -X(5,k)*sin(X(4,k)) X(5,k)*cos(X(4,k)) 0      0;...
       0                   0                  X(5,k) 0;...
       X(1,k)              X(2,k)             X(3,k) 1];
    x3D(:,:,k)=W'*xm;
end
```

Note that the homogeneous coordinates of the outline of Fig. 4.9, defined at the y-z plane, that can be used as an example of the array xm are stored (together with other data) in the MAT-file UPCoutlineData.mat.

Finally, the fourth of the homogeneous coordinates (fourth row of x3D) is simply discarded. The resulting outlines transformed into the 3D world coordinates are depicted in Fig. 4.15. ∎

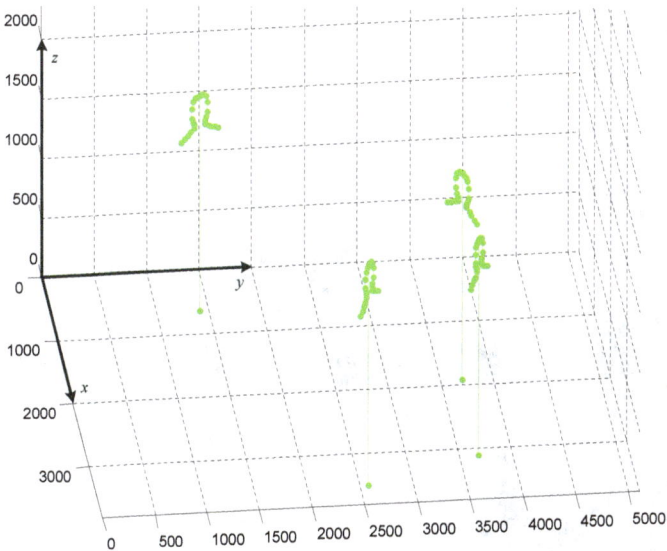

Fig. 4.15 Transformation of the upper torso outline model from the y-z plane into the 3D world coordinate system. Four outline instances are shown.

The projection of the $\mathbf{x}_{3D}^{(i)}$ onto the different camera views is done using the extrinsic and intrinsic camera parameters as discussed in section 4.2.

Example 4.10. Project the 3D outlines of example 4.9 onto the first four camera views of array I, given the camera parameters in the array of structures camData, both stored in the MAT-file UPCoutlineData.mat.

The camera calibration parameters are stored in MATLAB variables following the notation of the Camera Calibration Toolbox for MATLAB [Bouguet (2008)]. The extrinsic parameters are denoted by `Tc` and `Rc` (translation vector and rotation matrix respectively). The intrinsic parameters for the linear transformation are denoted `fc` (focal length vector), `cc` (principal point) and `alpha_c` (skew coefficient). The five nonlinear distortion parameters are stored in the 5×1 vector `kc`, ordered as $\left[k_r^{(2)}, k_r^{(4)}, k_t^{(1)}, k_t^{(2)}, k_r^{(6)} \right]^T$. Note that the 6-th order radial distortion coefficient $k_r^{(6)}$ comes fifth in `kc`. For compactness in notation, all these parameters are grouped in the given structure `camData`.

The pixel coordinates can be found using equations (4.1) to (4.6) and their MATLAB implementation by the function `world2cam()` discussed in example 4.1.

The resulting projected outlines are shown in Fig. 4.16 for all four camera views. ∎

Fig. 4.16   Projection of the four upper torso outlines from the 3D world coordinates of Fig. 4.15 onto the four camera views.

4.7 Moving objects

Most targets in visual tracking move. Humans especially, move all the time, even when they intend to be immobile. This motion can be very helpful in discriminating the target from the background.

There are two ways to use motion for visual tracking:

- Focus colour or outline matching to the motion areas only. This is a very low-level fusion of modalities, where image processing techniques are used to produce a motion mask. In the regions where there is very little motion, image pixels are discounted when colour histograms are calculated and edge pixels are removed when outline matches are sought. Such an approach is very sensitive to the definition of what very little motion actually is.
- Determine a motion likelihood $p\left(\mathbf{y}^{(\text{motion})}|\mathbf{x}\right)$ of the motion measurement $\mathbf{y}^{(\text{motion})}$ given the state $\mathbf{x}$. This is discussed in subsection 4.7.3.

There are two ways to measure motion:

- Short-term motion. This corresponds to finding all differences across frames in a video stream. The frame-by-frame difference (subsection 4.7.1) is a typical example.
- Long-term motion. This is utilised to learn a (usually adaptive) model for the background. Two such adaptive background estimation algorithms are outlined in subsection 4.7.2.

4.7.1 *Frame-by-frame difference*

The absolute difference of two frames highlights the differences between them. Ideally this will zero out all background, except the uncovered one, and will give large values to the moving foreground objects. Several factors do not allow such a clear distinction:

- Noise can create differences in background regions.
- Lighting (including shadows) can change, again creating differences in background regions.
- The colours of foreground moving objects can be similar to those of the surrounding background, making the absolute differences small.

- Uniformly coloured regions in the foreground objects result in no differences.

Hence the moving targets are manifested as differences moderately larger than other differences in the frame and only at the edges of the moving targets.

4.7.2 *Adaptive background estimation*

There are two approaches for adaptive background estimation: pixel-level modelling considers individual pixels, while appearance-based modelling considers the image as a whole.

Pixel-level modelling approaches are variants of Stauffer's adaptive background estimation algorithm [Stauffer and Grimson (2000)]. According to this, a model is built for every pixel in the frame. Gaussian Mixture Models (GMM) model the different colours every pixel can receive in a video sequence. The Gaussians are three-dimensional, corresponding to the red, green and blue components of the pixel colour. Their weight is proportional to the time a particular Gaussian models best the colour of the pixel. Hence the weight of a given Gaussian is increased as long as the colour of the pixel can be described by that Gaussian with higher probability than any other Gaussian in the GMM can, and that probability is above a threshold. Should it fall below the threshold, a new Gaussian is initiated in the GMM. If the maximum number of Gaussians is reached, the one with smallest weight is overwritten in the GMM. As a result, a map can be built in which every pixel is represented by the weight of the Gaussian from its GMM that best describes its current colour. This is the Pixel Persistence Map (PPM): regions of the map with large values correspond to pixels that have colours that appear there for a long time, hence are background. On the contrary, regions with small values correspond to pixels that have colours that appear there for a short time, hence are foreground. This is true as long as the foreground objects have colours distinct from the background. The problem with Stauffer's algorithm is in the case of foreground objects that stop moving. In its original implementation, targets that stop moving are learnt into the background. This happens as the weights of the Gaussians of the GMM of pixels describing the foreground colours and corresponding to immobile foreground objects increase with time. To alleviate this problem the learning rate can be spatiotemporally adapted by feeding back the areas in the image occupied by the targets and their velocities, as they are tracked by Kalman filters [Pnevmatikakis and Polymenakos (2006)]: the areas of slow moving targets are assigned smaller learning rates.

Appearance-based methods consider frames as a whole, attempting to capture in the model the correlation of neighbouring pixels. Principal Component Analysis (PCA) of a descriptive set of training frames is employed, yielding basis eigenimages, each with their associated eigenvalues. Eigenimages with large eigenvalues are the persistent ones, whose combination yields the background. There are two issues with this approach: the number of eigenimages to map as background and the on-line learning of changes in the background not described in the training images. The former is addressed empirically, while the latter involves on-line robust update of the eigenimages, as in [Skocaj *et al.* (2007)].

4.7.3 *Motion likelihood*

In [Pérez *et al.* (2004)], the motion likelihood is defined similarly to the colour one. The Bhattacharrya distance between the histogram $\mathbf{h_x}$ of a region $R_\mathbf{x}$ of the difference image, and a uniform reference one $\mathbf{h}_{\text{ref}}$ is utilised. The region depends on the state $\mathbf{x}$. It is rectangular centred at the position indicated by the state, while its size is larger than the size indicated by the state to include motion evidence from uncovered background. The likelihood for the motion measurement $\mathbf{y}^{(\text{motion})}$ is defined as :

$$p\left(\mathbf{y}^{(\text{motion})}|\mathbf{x}\right) \propto \exp\left(-D(\mathbf{h_x}, \mathbf{h}_{\text{ref}})/2\sigma^2_{\text{motion}}\right) \tag{4.24}$$

The rationale for the uniform reference histogram is that moving objects generate both large and small differences depending on the colours, patterns and motion. This generic reference histogram can be specialised for objects that are approximately circular disks, such as heads or torsos. Assume that the state is $\mathbf{x} = [\mathbf{x}_p, r]^T$, where $\mathbf{x}_p$ is the location on the image plane and r is the radius. The region is then a circular disk centred at $\mathbf{x}_p$ with radius $1.5r$. Contrary to using the intensity difference histogram inside $R_\mathbf{x}$ [Pérez *et al.* (2004)], a joint difference of intensity vs. distance from the centre histogram is utilised. Pixels at about the radius r, say in $[0.8r, 1.2r]$ are expected to exhibit large intensity difference, since they are at the circular object boundary and are affected by the motion of the objects. Pixels further away from the centre are expected to have small intensity difference, since they mostly belong to the background. Finally, pixels of smaller distance from the centre lie well within the object and their intensity difference depends on whether the object has uniform colour or not. Hence all intensity difference values are equally expected. The joint histogram is thus evaluated at three distance and four intensity difference regions.

4.8   Face detection

Faces play a paramount role in visual person tracking. Although quite variable in terms of pose, expression and illumination changes, they do not flex as much as other body parts and the geometry of their features does not change much across different people. This makes them easier to detect than other human parts.

The detection of faces looking at the camera and recorded with high resolution is considered a solved problem [Yang (2004)], that has already found its way to commercial implementations in photo cameras given typical portrait-taking distances (e.g. by Nikon [Nikon (2005)] or Fujifilm [Fujifilm (2006)]). On the other hand, face detection under arbitrary pose, orientation, scale, expression, imaging characteristics such as lighting and camera resolution) and face structural components (such as facial hair and glasses) is very difficult [Yang (2004)].

Many face detectors found in related literature utilise skin colour [Jones and Rehg (2002); Hsu *et al.* (2002); Jaffré and Crouzil (2003); Bradski (1998)], adding some constraints to the detected regions, and hence fall in the colour matching approaches discussed in section 4.5. The rest are split into those that attempt to find face features, and from them find the faces, and those that search directly for faces. The former are usually slower, and require adequate resolution for accurate feature detection. In multiple-person tracking applications the faces are small compared to the frame size; hence a good choice for a detector is the boosted cascade of simple classifiers, the Viola–Jones face detector [Viola and Jones (2001)].

The detected faces can be used in tracking in two different modes: either as contacts, or for forming a likelihood. Contacts indicate the existence of a face with absolute certainty. They are the native output of any face detector. The way they are obtained by means of the Viola–Jones face detector is discussed in section 4.8.1, while their validation to increase the detection certainty is discussed in section 4.8.2. The contacts can be used for target initialisation and tracker drift correction. On the other hand, all the detections in an image can be utilised to build a likelihood function of face existence. This second mode is suitable for integrating face measurements into the particle filtering framework and is discussed in section 4.8.3.

4.8.1 *Viola–Jones face detector*

The face detector based on boosted cascade of simple classifiers introduced by Viola and Jones employs multiple simple classifiers in a cascade. The rationale for the cascaded approach is that the performance of a single classifier is not suitable for object detection, since usually in order to get a high hit rate H (very close to unity) the classifier is penalised with a high false positive rate FP (about half). If now N such independent classifiers having similar hit rate H (very close to unity) and false positive rate FP (about half) are cascaded, then the resulting hit rate drops to H^N, which is still very close to unity, while the false positive rate drops dramatically to FP^N, which is very close to zero.

In every step down the cascade a decision is made, whether the image segment under consideration contains the face or not. The classifiers become more complex until the image segment is rejected or the end of the cascade is reached, whereupon the image segment is declared a face. The later classifiers in the cascade never check the same image segment again, if the area was previously rejected from a former classifier. In that way the algorithm works faster; background regions are discarded in early stages of the cascade and the algorithm focuses on more promising regions of the image.

Each classifier comprises features, checked against the image region for matching. We use the same Haar-like features (boxlets), introduced by Viola and Jones in [Viola and Jones (2001)]. Their black and white areas are matched to bright and dark areas of the image region under consideration. The classifier is matched against a portion of the image at a time, using a sliding window of variable size. The image is scanned first with a window of a minimum desired size, which is subsequently increased by a predetermined factor. This multiple-pass scanning of the image results in multiple candidate face bounding boxes of minor shifts and scale changes around the actual faces in the image. Should such bounding boxes exceed a predetermined number, they are merged into a face detection contact.

In order to select the classifiers for each cascade stage we are testing the cascade on a validation set. We are training each stage of the cascade separately. On each stage we keep on adding features, selected by the AdaBoost algorithm, up to the point that the target detection and false positives rates are met. We then keep on adding stages to the cascade up to the point that the overall target for false positive and detection rate is met. The approach has been extended to other features (e.g. parts in

[Schneiderman and Kanade (2004)]), or other boosting variants for training (e.g. Floatboost in [Li and Zhang (2004)]).

With regard to the implementation of the face detector, we follow the detector XML description introduced in OpenCV [Bradski *et al.* (2005)] and utilise the associated C function `cvHaarDetectObjects()` for invoking it. The detector described in `haarcascade_frontalface_default.xml` that comes with OpenCV is a very good general purpose frontal face detector, albeit not efficient with very small or non-frontal faces.

Should the need for a better detector arise, its stages can be trained using the provided OpenCV executable. Many positive samples (images of closely cropped faces) and negative samples (images with no human or animal face present) are needed. The positive samples are scaled to a standard size and are aligned by their eyes. If the faces are scaled to 12 pixels wide and 16 high (aspect ratio of 3/4) then typical computers with 2GB of memory can handle 9,000 positive and 18,000 negative samples. With regard to the particulars of the training algorithm, good choices are minimum feature size 0, 99.9% hit rate and 50% false alarm per cascade stage, horizontal and 40° tilted Haar-like features, non-symmetric faces, four splits and gentle AdaBoost learning [Bradski *et al.* (2005)].

4.8.2   *Face validation*

One approach to detect resolution-limited faces of varying pose, expression and illumination is to use multiple situation-specific detectors and fuse their decisions [Yang (2004); Schneiderman and Kanade (2004)]. Such an approach can be significantly time-consuming. Another approach is to make a generic detector less strict, allowing it to detect more poses, at the expense of more false positives. Such an approach needs a post-processing step, where image regions detected as faces are validated as such [Yang (2004)] using cues like texture for grey-scale images (e.g. the Distance From Face Space (DFFS) method [Moghaddam (2002)]), skin colour [Jones and Rehg (2002)] for colour images, motion for video sequences [Mikolajczyk *et al.* (2001); Karame *et al.* (2007)] and depth for stereo camera setups [Honda (2008)].

2D (image plane) validation involves discarding one of two detections that are overlapping on the image plane. This seldom removes actual faces, since partially occluded faces are usually not detected at all. The overlap is enumerated by the common area ratio of two candidate face regions. If two candidate face regions have bounding boxes BB_1 and BB_2, then their common area ratio is defined as

$$CAR = \frac{A[BB_1 \cap BB_2]}{\min\left(A[BB_1], A[BB_2]\right)} \tag{4.25}$$

where $A[BB]$ denotes the area of an image region bounded by the bounding box BB.

3D position validation involves the estimation of the 3D position of the face from a single calibrated camera [Katsarakis and Pnevmatikakis (2009)]. In order to do so the face width is assumed known and constant, which is approximately true for all adults. Referring back to Fig. 4.1, any pixel $\{x_p, y_p\}$ in an image plane corresponds to a line in the 3D world, starting at the origin of the camera coordinate system (the centre of projection) and passing through the pixel. Due to the depth uncertainty, the exact position on that line is not specified; all the points of that line are mapped onto the same pixel $\{x_p, y_p\}$ in the image plane. The points on that line are represented by the depth-normalised camera coordinates $\{x_n, y_n\}$. The actual 3D coordinates in the camera coordinate system can only be determined if the depth z_c, i.e the distance along the principal ray, is known. They are given by $\{x_c, y_c, z_c\} = \{x_n \cdot z_c, y_n \cdot z_c, z_c\}$. As discussed in section 4.2, the intrinsic camera parameters provide a non-linear mapping of $\{x_n, y_n\}$ onto $\{x_p, y_p\}$. The inverse mapping can be obtained numerically.

Every face on the image plane can be marked by its bounding box $\{x_f, y_f, w_f, h_f\}$, where $\{x_f, y_f\}$ are the top-left corner coordinates on the image plane, w_f is its width in pixels and h_f is its height. The image coordinates of the middle-left and middle-right pixels, corresponding to the temples on a frontal face, are then given by:

$$\{x_{pl}, y_p\} = \{x_f, y_f + h_f/2\} \tag{4.26}$$

$$\{x_{pr}, y_p\} = \{x_f + w_f, y_f + h_f/2\} \tag{4.27}$$

These two temple points are non-linearly mapped to depth-normalised camera coordinates $\{x_{nl}, y_n\}$ and $\{x_{nr}, y_n\}$ respectively using the intrinsic camera parameters. They are depicted as points C and A in Fig. 4.17. Note that in this figure the dimension y_c is omitted for clarity; only the optical axes z_c, across which uncertainty occurs, and x_c are drawn. The distance between the two points is the depth-normalised width of the face W_n, given by:

$$W_n = |x_{nl} - x_{nr}| \tag{4.28}$$

Human faces do not vary a lot in width. It is safe to assume that faces of adults are on average $W = 15$ cm wide. Hence, the actual temple camera coordinates, depicted as points D and B in Fig. 4.17, are on the extensions of lines OC and OA, at depth z, such that the distance between B and D, denoted as $|BD|$ is W. From the similarity of triangles OAC and OBD, we have:

$$\frac{W}{W_n} = \frac{|OD|}{|OC|} = \frac{|BD|}{|AC|} \tag{4.29}$$

Also, from the similarity of triangles OCE and ODF, we have:

$$\frac{|OD|}{|OC|} = \frac{|OF|}{|OE|} = \frac{z}{1} \tag{4.30}$$

From (4.29) and (4.30) we obtain the depth z as:

$$z = \frac{W}{W_n} \tag{4.31}$$

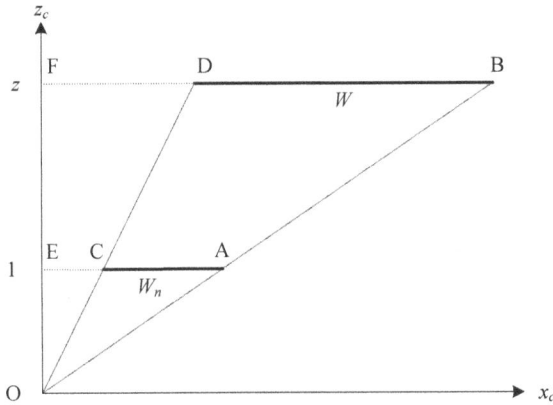

Fig. 4.17   Similar triangles used for depth estimate.

The face centre in 3D camera coordinates is then the midpoint between $\{x_{nl} \cdot z, y_n \cdot z, z\}$ and $\{x_{nr} \cdot z, y_n \cdot z, z\}$. This face centre can be linearly transformed from the camera coordinate system to the 3D world coordinate system using the extrinsic camera parameters, as discussed in section 4.2.

In the above derivation we have assumed a frontal face, facing the camera without any in- or out-of-plane rotations. These rotations can result in an approximately frontal face, the part of which that is visible to the camera is not temple to temple. In such cases the assumption of a face bounding box 15 cm wide can be crude. In addition to the effect of rotations, the actual width of a human face can vary around the average 15 cm. All these factors can somewhat degrade the 3D position estimate of the face.

To perform 3D position validation based on the 3D position estimation from a single calibrated camera two types of constraints are applied:

- Height constraints. The faces cannot be near the floor, nor at an unnatural height. We can ask for the height z_w to be $z_w \in [z_{min}, z_{max}]$ and discard all other detections.
- Floorplan constraints. The faces cannot be outside the walls of the monitored indoor space, if these are known. In this case, apart from z_w we constrain the two other coordinates to be $x_w \in [x_{min}, x_{max}]$ and $y_w \in [y_{min}, y_{max}]$. We discard all other detections.

Colour validation involves discarding faces based on the similarity between their colour histogram and a reference one, usually a skin colour histogram such as the one of Jones and Rehg [Jones and Rehg (2002)]. The histogram similarity is enumerated by the Bhattacharyya coefficient of section 4.5.

Pattern validation utilises a measure of similarity of the image pattern to a prototype frontal face. One such measure of similarity is the Distance From Face Space (DFFS) [Moghaddam (2002)], that is trained using a set of frontal faces. DFFS training involves the training of a face subspace projection matrix $\mathbf{W}$ using Principal Components Analysis (PCA) [Moghaddam (2002)]. The DFFS is the root-mean-square value of the elements of the difference between the original face vector $\mathbf{x}$ and the reconstructed projection of $\mathbf{x}$ into the subspace. The face vector is obtained by resizing any greyscale image region to a standard size and rearranging the resulting columns of grey levels into a vector of length N_x. The reconstruction difference vector is given by

$$\mathbf{d}_{proj} = \mathbf{x} - \left(\mathbf{W} \left(\mathbf{W}^T \left(\mathbf{x} - \overline{\mathbf{x}} \right) \right) + \overline{\mathbf{x}} \right) \tag{4.32}$$

where $\overline{\mathbf{x}}$ is the average face vector of the training set. Then DFFS is

$$DFFS = \sqrt{\frac{1}{N_x} \sum_{n=1}^{N_x} d_{proj}(n)^2} \tag{4.33}$$

where $d_{proj}(n)$ is the n-th element of the difference vector.

Note that DFFS validation is a good choice only when the faces that need validating are of the kind utilised to train the subspace projection matrix $\mathbf{W}$, i.e. they are frontal, and they are cropped by the detector in a manner similar to those in the training set.

4.8.3   Face likelihood

When the Viola–Jones frontal face detector is applied without merging of candidate face bounding boxes, then a multitude of bounding boxes are returned. Most of them are in large groups, with minor variations of their location and scale, bounding actual faces. Some others can be found in much smaller density around non-frontal faces and even around false alarms. All these N bounding boxes form the measurement vector $\mathbf{y}^{(\text{face})} = \left[[\mathbf{y}_p^{(1)}, y_w^{(1)}], \ldots, [\mathbf{y}_p^{(N)}, y_w^{(N)}] \right]^T$ where $\mathbf{y}_p^{(i)}$ is the two-dimensional position of the i-th bounding box and $y_w^{(i)}$ is its width. Note that the height needs not be specified, since the detector has the same aspect ratio for all the faces it reports.

Given the state vector $\mathbf{x} = [\mathbf{x}_p, x_w]^T$, the likelihood for the face measurement $\mathbf{y}^{(\text{face})}$ receives contributions from candidate face bounding boxes [Pnevmatikakis and Talantzis (2010)]. These contributions should be larger as the bounding box locations $\mathbf{y}_p^{(i)}$ approach the state location $\mathbf{x}_p$ and as their width $y_w^{(i)}$ approaches the state width x_w. Both goals are achieved by defining the likelihood as :

$$p\left(\mathbf{y}^{(\text{face})} | \mathbf{x} \right) = \sum_{i=1}^{N} \frac{w_i}{\left(\| \mathbf{x}_p - \mathbf{y}_p^{(i)} \|_2 / x_w^2 \right)^{K/2} + 1} \qquad (4.34)$$

Note that each term $\left(\left(\| \mathbf{x}_p - \mathbf{y}_p^{(i)} \|_2 / x_w^2 \right)^{K/2} + 1 \right)^{-1}$ makes a large contribution to the likelihood, close to unity if the distances of the state and the bounding box positions are close to zero. The exponent K governs how fast the contributions are attenuated as these distances increase. Good choices are $K = 1, 2$. The contributions are weighted in the likelihood summation by the weights w_i, defined as

$$w_i = \exp\left(- | x_w - y_w^{(i)} | / 2\sigma_w^2 \right) \qquad (4.35)$$

to penalise the contribution of candidate face bounding boxes that are quite different in width to x_w. The reason the difference in width is not included in the norm at the denominator of eq. (4.34) is that differences in bounding box locations are not comparable in scale to differences in bounding box widths. Putting width differences in the same norm as location differences would scale down the importance of the former relative to the latter. Also note that the weights in eq. (4.35) are not scaled to sum to unity. This is chosen so that multiple similar detections would increase the likelihood compared to a single detection. A necessary penalty to pay is that the likelihood values in eq. (4.34) are not bounded by unity.

The face likelihood obtained by eq. (4.34) is a three-dimensional function, one for each dimension of the state. An example is given in Fig. 4.18. Evaluating the likelihood for different state locations $\mathbf{x}_p$ results in the values of the different pixels on the likelihood plane. Evaluating for the different state widths $y_w^{(i)}$ results to the different likelihood planes. Note how the likelihood peaks in the vicinity of the bounding boxes and does more so as additional bounding boxes are located nearby. Also note the effect of the weights in (4.35) in selecting faces of the required widths.

Fig. 4.18 Image and associated face likelihoods according to eq. (4.34), evaluated across all the image plane for three face widths.

4.9   Visual tracking systems

Some possible measurement cues to be used in visual tracking have been detailed in sections 4.2 to 4.8. These measurements are now to be applied in tracking systems.

From an algorithmic point of view, a sequential Bayesian tracker is required to update the previous state to the current one. For Kalman filters, the state is represented by a vector, while for particle filters, it is the set of particles and their weights.

A tracking system on the other hand needs to initialise, maintain and eventually terminate multiple targets. Multiple visual targets in this book are handled by assigning an independent tracker to each of them.

In this section we describe a loop of the tracking system, i.e. the operations happening with every new frame being processed. The tracking algorithm itself is only a part of the complete system.

For every new frame arriving to be processed by the tracking system a coarse detection is first attempted. This is done by measuring for cues, usually multiple ones, across a grid spanning the space of interest and the different scales. This grid is quite coarse for fast operation. The scale is restricted by the recent target size statistics maintained by the system. The process yields the coarse evidence, termed contacts. If no contacts exist, the target size statistics are reset to allow for broader scale search in the next frame. This way abrupt changes in the zoom are handled.

The contacts are first used to update target size statistics. This way progressive changes in the zoom are accounted for. Note that for tracking systems operating with fixed cameras, there is no need to maintain target size statistics since the zoom does not vary.

Subsequently the contacts are associated with existing targets based on the Euclidean distance of their coordinates, normalised by the width of the target. Only pairs not exceeding a maximum association distance are associated. An optimal greedy algorithm, the Hungarian (or Munkres) algorithm [Blackman and Popoli (1999)] is used for the association. The algorithm minimises the overall distance between the targets and the contacts. An excellent implementation of the algorithm for MATLAB can be found in [Buehren (2009)]. The association process is detailed in example 4.11.

Example 4.11. Assume that the detection process has reported K_y contacts as vectors of two location elements $\mathbf{y}_{k_y} = [x_y, y_y]^{\mathrm{T}}$ and there are currently K_x targets as vectors of two location elements and one width element $\mathbf{x}_{k_x} = [x_x, y_x, w_x]^{\mathrm{T}}$. In MATLAB, the contacts are concatenated into matrix $\mathbf{Y}$ of size $2 \times K_y$ and the target states into matrix $\mathbf{X}$ of size $3 \times K_x$. The matrix $\mathbf{D}$ of size $Kx \times K_y$ of normalised Euclidean distances of the targets from the contacts is then obtained as:

```
D=zeros(Kx,Ky);
for kx=1:Kx
    for ky=1:Ky
        t=Y(:,ky)-X(1:2,kx);
        D(kx,ky)=sqrt(t'*t)/X(3,kx);
    end
end
```

In order not to attempt to associate the contacts that are far from the evidence with the closest one, distances exceeding a threshold d_{thr} are set to infinity:

```
D(D>dthr)=Inf;
```

The association is performed using the implementation of the Hungarian algorithm by the MATLAB function `assignmentoptimal()` from [Buehren (2009)]:

```
a=assignmentoptimal(D);
```

The vector $\mathbf{a}$ is of length K_x and its element $\mathbf{a}_{k_x}$ contains the index of the contact associated with the k_x-th target. If $\mathbf{a}_{k_x} = 0$, then the k_x-th target is not associated with any contact. Inversely, the indices of the unassociated contacts are given by `setdiff(1:ny,a)`. ∎

Unassociated contacts are used for target initialisation, if they pass stricter validation criteria. This also involves the training of target models for some of the measurement cues.

All targets except those just initialised are tracked using a tracking algorithm from chapter 2. Tracking also involves updating target models for the measurement cues utilised. The models of the targets are updated if the respective model matching quality is above a threshold, indicating trusted tracks. These thresholds have to be strict, otherwise tracks drifting away from their targets will have their models updated based on the wrong part of the frame. The nature of the update depends on the measurement cue. For example, the model used for face detection needs hours to train, hence it is not easy to update. A fixed generic one is used instead, trained as discussed in section 4.8.1. The model for outline matching can change its appearance beyond that allowed by the deformations indicated by the state. For example the shoulders in an upper torso model can be included in the outline model, or it can deteriorate to a head model. The decision is based on the matching quality with and without shoulders. Model deterioration to head-only is useful for profile bodies. The colour model is updated every time the face detection likelihood is very high, indicating a fine face detection that matches nicely with the tracked face. The update is based on combining the reference histogram up to the previous frame with the reference histogram from the current frame using a learning rate.

During tracking the model matching qualities for all cues are monitored, and the number of consecutive frames where these qualities are too low is counted. The quality thresholds can vary linearly between a strict and a loose one, as a function of the age of the target. In this way younger targets are handled differently to older ones. When the target models do not match well with the evidence collected it is an indication of change in appearance, tracker failure or disappearance of the target from the scene. Targets with momentarily low matching qualities might not be reported, while those for which the low quality persists should be terminated.

Both the existence of coarse evidence associated with the target and the measurement of very high matching qualities can indicate a contact, i.e. there is sufficient measurement evidence supporting the existence of the target. A recent history of few contacts leads to not reporting the target, while one of too few contacts leads to target termination.

Target termination is also based on particle spread. A large spread indicates particles with no reason to lock in a specific frame location, hence they are spread apart by means of the random components of the object model. Too large spread indicates lack of target lock in a different way than the model matching quality. In cases of measurement ambiguity (e.g. background colours similar to those of the target), the model matching quality can be high in a whole frame region, resulting in high quality and high spread.

4.10 Deterministic tracking

Although not strictly in the scope of this book, for the sake of completeness two deterministic visual tracking algorithms are presented in the following paragraphs. They are the template matching and the Mean-Shift ones.

4.10.1 *Template matching*

Template matching tracking algorithms track an image patch T (the template) comprising a set of K pixels $\mathbf{p}_k$, $k = 1 \ldots K$ extracted at time $n - 1$ in the current frame I at time n. During the single time step the template can undergo any transformation $\mathbf{W}(\mathbf{p}; \mathbf{x})$, where $\mathbf{x}$ is the vector of transformation parameters to be determined.

The determination of the transformation parameters is done by minimising the discrepancy between the template $T(\mathbf{p})$ and the image region transformed back into the coordinate frame of the template, $I(\mathbf{W}(\mathbf{p}; \mathbf{x}))$. One way to minimise the discrepancy is to minimise the sum of the squared errors for all $\mathbf{p}_k$, hence $\mathbf{x}$ is obtained as:

$$\mathbf{x} = \underset{\mathbf{x}}{\operatorname{argmin}} \left(\sum_{k=1}^{K} \left[I\left(\mathbf{W}(\mathbf{p}_k; \mathbf{x})\right) - T(\mathbf{p}_k) \right]^2 \right) \tag{4.36}$$

In the case of a transformation $\mathbf{W}$ comprising only shifts along the two image axes, the minimisation in (4.36) can be carried out by employing a search around the current position of the target i.e. for some pixels along each image dimension. For rectangular templates, that search becomes a block matching search equivalent to the one in video compression [Tekalp (1995)].

In the case of the more general affine or projective transformations, where the parameters $\mathbf{x}$ are not all discrete, grid search becomes very inefficient. The solution in this case is the Lucas–Kanade algorithm [Lucas and Kanade (1981)], that iteratively estimates the parameters. Assuming some parameter estimation $\mathbf{x}_{i-1}$ at iteration $i - 1$, the estimation for the next iteration is found incrementally as $\mathbf{x}_i = \mathbf{x}_{i-1} + \Delta\mathbf{x}_i$. In order to do so the increment is found as:

$$\Delta\mathbf{x}_i = \underset{\Delta\mathbf{x}_i}{\operatorname{argmin}} \left(\sum_{k=1}^{K} \left[I\left(\mathbf{W}(\mathbf{p}_k; \mathbf{x}_{i-1} + \Delta\mathbf{x}_i)\right) - T(\mathbf{p}_k) \right]^2 \right) \tag{4.37}$$

The minimisation is achieved by differentiating with respect to $\Delta\mathbf{x}_i$, as in [Lucas and Kanade (1981)]. The process is repeated until convergence, indicated by a termination criterion on the minimum magnitude of $\Delta\mathbf{x}_i$, is satisfied.

4.10.2 *Mean-Shift*

Mean-Shift (and its Continuously Adaptive version) is a non-parametric iterative technique for finding the mode of a target probability distribution

[Jaffré and Crouzil (2003); Comaniciu *et al.* (2000)], while dynamically adjusting the parameters of the distribution [Bradski (1998)]. It can be applied to visual tracking to update the state vector at the previous time instant $\mathbf{x}_{n-1}$ into the one at the current time instant $\mathbf{x}_n$ in an iterative manner as follows.

Assume a target of constant aspect ratio a. The state vector comprises the 2D position (x,y) of the target centre and its width W. The height is simply $a \cdot W$. A subscript is added in the notation to indicate current (n) or previous $(n-1)$ time instant. Given the current frame, a binary image B can be constructed by using measurement cues to validate how well the pixels match a measurable target property. The target centroid at the current instant $[x_n, y_n]^T$ is iteratively approximated from its position at the previous instant $[x^{n-1}, y^{n-1}]^T$. The iteration is initiated by setting:

$$\left[x_n^{(0)}, y_n^{(0)}\right]^T = [x_{n-1}, y_{n-1}]^T \tag{4.38}$$

The centroid for instant n and the i-th iteration step is then given by:

$$\left[x_n^{(i)}, y_n^{(i)}\right]^T = \frac{\left[M_{10}^{(i)}, M_{01}^{(i)}\right]^T}{M_{00}^{(i)}} \tag{4.39}$$

where $M_{00}^{(i)}$ is the zeroth-order moment for the i-th iteration step computed around $[x_n^{(i-1)}, y_n^{(i-1)}]^T$:

$$M_{00}^{(i)} = \sum_{x=x_n^{(i-1)}-W_{n-1}(1/2-m)}^{x_n^{(i-1)}+W_{n-1}(1/2+m)} \sum_{y=y_n^{(i-1)}-a\cdot W_{n-1}(1/2-m)}^{y_n^{(i-1)}+a\cdot W_{n-1}(1/2+m)} B(x,y) \tag{4.40}$$

and $M_{10}^{(i)}$, $M_{01}^{(i)}$ are the first-order moments for the i-th iteration step computed around $[x_n^{(i-1)}, y_n^{(i-1)}]^T$:

$$M_{10}^{(i)} = \sum_{x=x_n^{(i-1)}-W_{n-1}(1/2-m)}^{x_n^{(i-1)}+W_{n-1}(1/2+m)} \sum_{y=y_n^{(i-1)}-a\cdot W_{n-1}(1/2-m)}^{y_n^{(i-1)}+a\cdot W_{n-1}(1/2+m)} x \cdot B(x,y) \tag{4.41}$$

and:

$$M_{01}^{(i)} = \sum_{x=x_n^{(i-1)}-W_{n-1}(1/2-m)}^{x_n^{(i-1)}+W_{n-1}(1/2+m)} \sum_{y=y_n^{(i-1)}-a\cdot W_{n-1}(1/2-m)}^{y_n^{(i-1)}+a\cdot W_{n-1}(1/2+m)} y \cdot B(x,y) \tag{4.42}$$

Note that in equations (4.40),(4.41) and (4.42) m is an expansion margin around the target estimation at the previous iteration step. Convergence is achieved when the shift between the centroids estimated at the $i-1$ and

i iteration steps is less than a pixel. After convergence at iteration i we estimate the updated width W_n at instant n by approximating the area of the target bounding box as the converged zeroth-order moment, resulting in:

$$W_n = \sqrt{a\, M_{00}^{(i)}} \tag{4.43}$$

4.11   Stochastic tracking

4.11.1   *Kalman filter*

The Kalman filter, as an exact solution to the recursive Bayesian filtering problem for the case of linear Gaussian dynamic models, was introduced in section 2.4.2. In this section examples of visual tracking systems employing Kalman filters are detailed, starting with a system for tracking fingertips on a multi-touch surface and continuing with a system for tracking pedestrians and vehicles in outdoor scenarios.

4.11.1.1   *Fingertip tracker*

Kalman filters are used to track fingertips as they are recorded by a camera. The camera is part of a multi-touch surface that serves to attenuate background clutter, giving very clear measurement contacts.

The hardware setup for a multi-touch surface is as follows [Petsatodis *et al.* (2009)]: an acrylic panel is placed on top of the TFT panel, the edges of which are illuminated by IR-Light Emitting Diodes. Due to the Frustrated Total Internal Reflection effect, a finger touch on the surface of the acrylic panel generates lighting blobs. This is due to the fact that the refractive index of skin is higher than that of air. The position of blobs that manage to penetrate the TFT panel is captured by a near-infrared (NIR) camera through an ultraviolet and visual cut optical filter.

The images captured from the NIR camera are binarised using a constant threshold, due to the absence of adverse illumination effects ensured by the hardware design. The resulting active pixels are grouped into objects using 4-way connectivity. These objects form the pool of two types of contacts: those used for track initialisation, and those used at the measurement update stage of the Kalman tracker(s). One Kalman tracker is employed per target, tracking it independently from the rest. The initialisation contacts are selected based on strict size and aspect ratio criteria. This minimises false positives, at the risk of delaying the initialisation of a track in two cases: when originally a fingertip touches the surface too lightly (small object size) or when more than the fingertip touches the surface producing elongated objects (large object aspect ratio). The measurement update contacts are selected using more relaxed size and aspect ratio criteria. This is safe as these contacts do not generate false positives but on the other hand facilitate the maintenance of an existing track should the fingertip temporarily produce erratic objects on the binary frames.

The states according to the Kalman filtering formulation are represented by their mean vectors and covariance matrices. For the finger tracker, the states are four element vectors, comprising 2D position (x,y) and velocity (v_x,v_y). During the tracking cycle, every Kalman tracker first updates the state of all targets based on a constant velocity object model [Blackman and Popoli (1999)], as discussed in example 2.2. The linear term of the dynamic model is constant and is given by (2.26). Similarly the covariance matrix of the process noise is given by (2.27).

To implement the state update stage of the Kalman filter in MATLAB, the mean vectors and covariance matrices representing the states are concatenated into a matrix xkk and three-dimensional array Ckk. The last dimension in both these MATLAB variables is the index of the target. The state update described by (2.18) and (2.19) then yields the one-step predictions for the mean vector (tt x01) and the covariance matrix (tt C01) using the following MATLAB code:

```
x01=F*xkk;
for k=1:size(xkk,2)
    C01(:,:,k)=F*Ckk(:,:,k)*F'+Q;
end
```

Please note that, unlike example 2.2 where the last dimension in xkk, Ckk, x01, C01 corresponds to time, here it corresponds to target. Time history is not stored, since it is not necessary.

Following the state update step, all the updated states are associated with measurement update contacts using the Hungarian (or Munkres) algorithm [Blackman and Popoli (1999)], as detailed in example 4.11. The only difference is that the distance is a simple Euclidean, not a normalised, one. The associated contacts, in the form of the measurement vectors $[x, y]^T$, are used for the measurement update stage of the Kalman trackers. The measurement model (see (2.15)) comprises the constant across time linear ($\mathbf{H}$) and noise ($\mathbf{R}$) terms given by (2.24) and (2.25) respectively, as discussed in example 2.2. The variance of each position element σ_R^2 is $\frac{\Delta t^2}{10}$.

The measurement update stage of the Kalman filter is then implemented for every target, using the following MATLAB code. Firstly, the covariance of the innovation term S, the Kalman gain K and the updated state covariance Ckk are calculated using (2.20), (2.21) and (2.23) respectively. Then the vector of associations a returned from the association algorithm is examined. If the target has an associated contact, then the state mean xkk is updated based on (2.22). If this is not the case, then the state mean is just the one-step prediction x01 already calculated.

```
for k=1:size(xkk,2)
    S=H*C01(:,:,k)*H'+R;
    K=C01(:,:,k)*H'*inv(S);
    Ckk(:,:,k)=C01(:,:,k)-K*H*C01(:,:,k);
    xkk(:,k)=x01(:,k);
    if a(k)>0
        % Contact a(k) is associated with this state
        xkk(:,k)=xkk(:,k)+K*(y(:,a(k))-H*x01(:,k));
    end
end
```

Any states without contacts associated with them are kept active using the constant velocity updates for up to a third of a second. If in that interval there is no association, the track is terminated. The associated measurement update contacts are then removed from the pool of objects and the initialisation contacts are selected from the remaining objects using stricter criteria. These initialise new Kalman filters with a unique track ID.

At every frame, the fingertip tracking system reports the IDs and the positions of all active tracks (fingertips). Typical operation of the finger tracking system is shown in Fig. 4.19.

4.11.1.2 *Body tracker*

Determining the contacts in an visual body tracker, especially for outdoor scenarios , is much more challenging than for the NIR images of the multitouch surface. In order to do so, the trackers control the measurement process [Pnevmatikakis and Polymenakos (2006)]: the adaptive foreground segmentation scheme outlined in section 4.7.2 is used. Stauffer's adaptive foreground segmentation algorithm [Stauffer and Grimson (2000)] is modified to accommodate a learning rate that is spatiotemporally adapted: image regions corresponding to targets will have learning rates proportional to the speeds of the targets. In order to achieve this, a feedback loop is

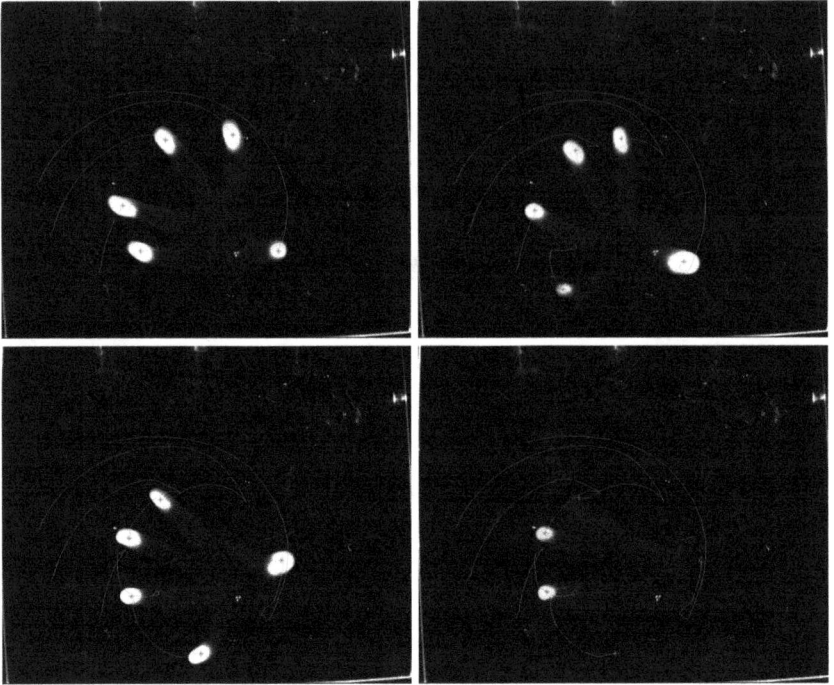

Fig. 4.19   Typical operation of the tracking system for five simultaneous contacts.

created with Stauffer's algorithm, a target management system and a tracker per target, as shown in Fig. 4.20.

The states comprise 2D plane position information and 2D velocity. A constant velocity object model is used, together with Gaussian process and measurement noises. The measurement model is a linear mapping of the position part of the state to the image plane. Hence the trackers can be Kalman filters to avoid unnecessary complexity of the particle filters.

Since the position part of the state points at a rectangle on the image and the velocity governs the value of the learning rate in these rectangles, the Kalman filters effectively modify the measurement process. The learning rate of fast targets does not change from the large overall one, since these targets are not in danger of fading into the background. As the speed reduces, then pixels belonging to the target tend to remain the same for longer periods of time. The learning rate is then reduced to slow down adaptation.

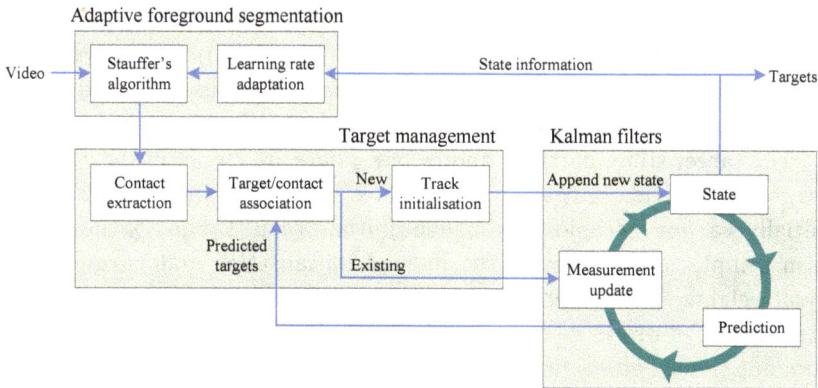

Fig. 4.20   Outdoor tracking system with the Kalman filters controlling the measurement process.

4.11.2   *Colour-based particle filter face tracker*

This section and the following ones give examples of particle filter trackers for tracking faces at the 2D plane. The goal is to demonstrate the different options for particle filtering in a realistic setup instead of in the 1D cases in Chapter 2. By doing so, the issue of optimum proposal distributions is further explored, leading to a measurement-assisted version.

The state comprises the 2D position of the face centre, the face width and height: $\mathbf{x} = [x, y, w, h]^{\mathrm{T}}$. The mapping of the state on an image region $\mathbf{x} \rightarrow R_{\mathbf{x}}$ creates a rectangle with top-left corner $[x - w/2, y - h/2]$ and bottom-right corner $[x + w/2, y + h/2]$. All measurements occur in this rectangle.

The face trackers built in this and the following sections operate on a sequence where the single face present is annotated. This allows for actual scoring the accuracy of the tracker. The target moves a lot, with both scale and pose changes. The single face target is manually initialised in the first frame. The use of the sequence and its annotations for tracking a target is described in example 4.12

Example 4.12. This example details how to process an annotated video by a single target tracker. The target is present throughout the video and is manually initialised at the first frame.

The particle filter tracker has numerous parameters that are set at the initialisation stage in the first six lines of code. Their meaning is examined in examples 4.13 and 4.14 where the particle filters are discussed.

To load frames from video (AVI) files in MATLAB, a third party open source toolbox, the VideoIO toolbox [Dalley (2009)], is used. A video object is generated using `vr=videoReader(<filename>)`. Subsequent frames are read using `I=getnext(vr)` and the video file is closed using `vr=close(vr)`.

The target state, particles and colour model are stored in the structure `target` with fields `target.xo`, `target.x`, `target.w` and `target.Hface`. Initially `target.xo` holds the manual initialisation, `target.x` the Np uniform samples drawn around the manual initialisation and `target.w` the equal weights. The target colour model is initialised by calling the function `histogram()` introduced in example 4.3 and is stored in `target.Hface`. It does not change after initialisation.

After initialisation, in a loop spanning all the frames, the tracker code is run (the MATLAB script named `SIR.m` in the following code) and then the results are scored. The annotations contain the position of both eyes and the centre of the mouth. From those, using simple face geometry assumptions a face bounding box can be derived, which might not be perfect in extreme poses but is accurate enough for scoring purposes. The annotated face bounding boxes are stored in the array `x_a` in the MAT file `annotations.mat` as centre coordinates, width and height. The tracking accuracy is measured with the position error `posErr` which is the offset between the tracked and annotated centres and with the size error `sizErr` which is the offset between the tracked and annotated widths and heights. The offsets are calculated relative to the annotated width and height of the face. A final figure of merit that shows the track quality (even without annotations) is the spread of the particle samples. To calculate this, the standard deviations of the dimensions of the samples are normalised: the horizontal position and width standard deviations are divided by the face width, and the vertical position and height standard deviations by the face height. The spread is defined as the maximum of these normalised standard deviations. Note that if there are no annotations the standard deviations can be normalised by the estimated width and height.

All the above are implemented by the following MATLAB code:

```
Nh=32;          % Quantisation levels per colour component for histogram
Np=100;         % Number of particles
s2c=.25;        % Colour likelihood variance
Lambda0=[.18 .18 .02 .02];% Random walk std relative to target size
bRW=0.75;       % Proposal distribution random walk over grid weight
fgrid=6;        % Fraction of random walk variance for grid points
bu=0.05;        % Object model uniform over random walk weight
vr=videoReader('apneFace.avi');
```

```
I=getnext(vr);
x=[910;520;280;280]; % Manual initialisation
bb=[x(1)-ceil(x(3)/2)+1;x(2)-ceil(x(4)/2)+1;x(3:4)];
target.Hface=histogram(I(bb(2):bb(2)+bb(4)-1,bb(1):bb(1)+bb(3)-1,:),Nh);
% Uniform initialisation of particle samples
Lambda=Lambda0*sqrt(x(3)*x(4));
per=rand(4,Np);
for i1=1:4
    per(i1,:)=per(i1,:)*Lambda(i1)-Lambda(i1)/2;
end
target.x=repmat(x,1,Np)+per;
target.xo=x;
target.w=ones(1,Np)/Np;
load annotations % Load manual annotations for scoring
posErr=zeros(get(vr,'numFrames')-1,1);
match=posErr;sizErr=posErr;
pOM=1;Ng=0;
for frame=2:get(vr,'numFrames')
    I=getnext(vr);
    x=target.x;
    w=target.w;
    xo=target.xo;
% Tracking code follows
    SIR
    xo=x*w';
% Scoring
    s=sqrt(prod(x_a(frame,3:4)));
    spread(frame-1)=max(std(x(1:4,:),[],2)'/s);
    posErr(frame-1)=sqrt(sum(((target.xo(1:2)'-x_a(frame,1:2))/s).^2));
    sizErr(frame-1)=sqrt(sum(((target.xo(3:4)'-x_a(frame,3:4))/s).^2));
    target.xo=xo;
end
vr=close(vr);
```

■

The first face tracker employs colour measurements. Although it is expected that the SIS tracker will not work at all due to the degeneration problem discussed in section 2.5.3, it is included here for comparison purposes. As expected, the object model keeps spreading the particles to unnecessary subspaces of the state-space, where they do not have any significant weight, but most importantly do not allow enough particles to explore the relevant subspace of the state-space. The performance of the SIS tracker is summarised in Table 4.2. Since the RMS position and size errors relative to the face size for the complete sequence are far above unity it is evident that the target is lost for the majority of the frames. Furthermore, the large particle spread designates that they are spread in the entire state-space.

Table 4.2   Particle filter face tracker: comparison of SIS versus SIR.

PF type	State estimation	Position error (%)	Size error (%)	Spread (%)
SIS	MMSE	429	205	237
SIR	MMSE	54.4	92.6	96.5
SIR	MAP	20.7	31.7	82.9

Table 4.3   SIR face tracker with bimodal (Gaussian and uniform) proposal distribution: effect of the number of particles employed.

Particles	Position error (%)	Size error (%)	Spread (%)
25	31.6	68.0	92.4
50	25.2	45.6	86.1
100	20.7	31.7	82.9
200	17.6	25.6	81.2

Resampling is introduced to the SIS face tracker, resulting in the SIR face tracker. As expected from the discussion in section 2.5.5, the particle filter tracker now actually works. The various options explored with the SIR tracker are discussed below.

Firstly, MAP versus MMSE state estimation is considered. The MMSE estimation is the mean of the particle samples, weighted by the particle weights. The MAP estimation is simply the particle sample with the largest particle weight. The two approaches converge when a few similar particles are much more weighted than the rest, but in cases of measurement ambiguity this is not the case. In this particular sequence MAP estimation greatly outperforms the MMSE one, since at the intervals of high motion the majority of the particles are left trailing the target. Although the weight of the few on the target is larger, the weighted average is nevertheless biased towards the previous location. The performance of the SIR tracker with either MAP or MMSE state estimation is also summarised in Table 4.2.

Secondly, the effect of the number of particles is investigated. As already demonstrated in examples 2.4 (Fig. 2.8) and 2.7 (Fig. 2.13), involving more particles leads to improved state estimation. This is also the case with the colour-based particle filter face tracker, as shown by the results in Table 4.3. As the number of particles increases, the performance gain is diminished, while the computational cost becomes huge.

Table 4.4   SIR face tracker with bimodal (Gaussian and uniform) proposal distribution: effect of position standard deviation relative to the face size.

Standard deviation	Position error (%)	Size error (%)	Spread (%)
0.12	23.6	37.3	81.4
0.15	22.3	33.3	83.8
0.18	20.7	31.7	82.9
0.21	20.7	34.5	84.4

Next, the effect of the variance in the Gaussian random walk component is considered. This variance allows the particles to spread. Larger values are needed for more rapid changes of the state of the target. The size of the face in the test sequence does not change very rapidly, but its position does change in the scenes of fast motion. On the other hand there are scenes with very little motion, where particle spreading results in unnecessary exploration of the state-space. The results are shown in Table 4.4. In the given sequence, it appears that changing the position standard deviation above 15% of the face size in every dimension has little effect, whilst 18% is optimum. Increasing the position standard deviations leads to only a minor increase in particle spreads; most of the variance is absorbed by resampling towards good particles.

The effect of the colour likelihood variance is to set how forgiving the measurement process is to colour variations from the model. Small variances will render the likelihood of patches with moderate colour changes from the reference to be negligible, requiring very good matches and concentrating the particles in those areas. On the other hand if the target actually changes appearance, this will lead to particle spread and target loss. In this particular short example, where the action is located indoors, the target does not change colour significantly, so very small variances seem to work better, but are by no means guaranteed to do so in more complex situations. The results are shown in Table 4.5.

Finally, the effect of the chosen proposal distribution and object model is considered. As discussed in section 2.5.1, a good choice in the absence of motion information for the target is the Gaussian random walk. On the other hand, target recovery in cases of loss is aided by a uniform component, spanning a much larger subspace of the state-space. Indeed, adding the uniform component with a relative weight $\beta_u = 0.05$ in (2.51) leads to a considerable reduction in the position and size errors, as shown in the first two rows of Table 4.6.

Table 4.5   SIR face tracker with bimodal (Gaussian and uniform) proposal distribution: effect of colour likelihood variance.

Colour likelihood variance	Position error (%)	Size error (%)	Spread (%)
2^{-1}	24.6	43.4	130
2^{-2}	20.7	31.7	82.9
2^{-3}	18.4	27.9	59.1
2^{-4}	17.0	24.5	49.3
2^{-5}	16.4	25.0	45.1

Table 4.6   SIR face tracker: effect of different proposal distributions.

Proposal	Position error (%)	Size error (%)	Spread (%)
Gaussian	24.6	45.8	65.6
Gaussian + Uniform	20.7	31.7	82.9
Gaussian + Grid	15.5	27.6	32.5

The SIR filter used for the derivations of all the results in this section is given in example 4.13.

Example 4.13. Implementation of the SIR particle filter face tracker with colour measurements and Gaussian plus uniform proposal.

To sample the proposal distribution, the Gaussian and uniform samples are first generated and one of the two is selected based on the relative weight `bu`, as discussed in example 2.5. The covariance matrix of the Gaussian random walk component utilises the standard deviations per component relative to target size given by `Lambda0` in its diagonal, but also adds some covariance between the width and height components, not allowing them to change completely independently.

The colour matching likelihood is then evaluated per particle. Each particle defines a bounding box in the image where the face is cropped and its histogram is constructed. It is compared to the reference in `target.Hface`, as discussed in example 4.4. The colour likelihood variance is `s2c`.

The SIR particle filter concludes an iteration with resampling, as discussed in example 2.7.

The MATLAB code for the particle filter is given below. The main difference from the code segments given throughout the examples in Chapter 2 is that the state is now multi-dimensional and there are several checks that the particles do not lead to measurements outside the image boundaries.

```
% Sample proposal distribution for updated particles
%
% 1. Gaussian (random walk) component
% Covariance matrix for Gaussian random walk
Lambda=Lambda0*sqrt(target.xo(3)*target.xo(4));
lamdac=sqrt(Lambda(3)*Lambda(4));
C=diag(Lambda.^2);C(3,4)=lamdac;C(4,3)=lamdac;
x=x+sqrt(C)*randn(size(x,1),Np);
% 2. Uniform component
xu=diag([size(I,2)/2 size(I,1)/2 xo(3)/20 xo(4)/20]) *(rand(size(x,1),Np)-.5);
xu=xu+repmat(xo,1,Np);
idx=find(rand(1,Np)<bu);
x(:,idx)=xu(:,idx);
% Make sure centre remains inside image
x(x<1)=1;
x(1,x(1,:)>size(I,2))=size(I,2);
x(2,x(2,:)>size(I,1))=size(I,1);
% Weight update using colour for position only state update
%
% 1. Evaluate colour matching likelihood model
pOM=zeros(1,Np);
for i1=1:Np
    bb=round([x(1,i1)-ceil(x(3,i1)/2)+1;x(2,i1)-ceil(x(4,i1)/2)+1;x(3:4,i1)]);
    bb(bb<1)=1;
    if bb(1)+bb(3)>size(I,2)
        bb(3)=size(I,2)-bb(1)+1;
    end
    if bb(2)+bb(4)>size(I,1)
        bb(4)=size(I,1)-bb(2)+1;
    end
    if bb(3)>5 && bb(4)>5
        Hface=histogram(I(bb(2):bb(2)+bb(4)-1,bb(1):bb(1)+bb(3)-1,:),Nh);
        pOM(i1)=exp(-(1-sum(sqrt(Hface.*target.Hface)))/(2*s2c));
    end
end
% 2. New weights
w=w.*pOM;
w=w/sum(w);
% Resample
%
c=cumsum(w);
u=rand(1)/Np;
xr=zeros(size(x));
j=1;
for i=1:Np
    while c(j)<u
        j=j+1;
    end
    xr(:,i)=x(:,j);
    u=u+1/Np;
```

```
end
target.x=xr;
target.w=repmat(1/Np,1,Np);
```

Note that for the SIS particle filter, the resampling section of the code should be omitted.   ■

4.11.3   *Measurement-assisted proposal distribution*

Blindly aiming at lock recovery in cases of target loss with a uniform component in the proposal distribution and the object model has two drawbacks:

- Subspaces of the state-space are randomly (uniformly) searched for evidence of the target. Most of these uniformly distributed particles are finding nothing.
- Particles are dedicated to the uniform component even when there is no target lock loss.

It is therefore dangerous to allocate too many of the particles to the uniform component, hence its relative weight β_u in (2.51) needs to be small and lock recovery is not guaranteed.

Proposal distribution design options have been described in section 2.5.4. Here, instead of utilising the Gaussian plus uniform mixture, we build a measurement-assisted proposal distribution by following the approach of [Pérez *et al.* (2004)]: the proposal distribution has the Gaussian random walk component with relative weight β_{RW}, plus the contribution of the measurement model in the form of a sum of Gaussian densities with relative weight $(1 - \beta_{RW})$. To establish the contribution of the measurement model, a grid $\mathbf{x}_g$ around all the particles is searched for the n_g locations $\left\{ \mathbf{x}_g^{(k)} \right\}_{k=1}^{n_g}$ of good measurement match.

At this point we deviate from the approach of [Pérez *et al.* (2004)] in two ways:

- We use this measurement-assisted proposal for the object model as well, hence the resulting particle filter tracker remains an SIR one.
- We do not use a constant threshold to define good measurement match. Instead, we use a fraction τ_g of the best match obtained at the previous time instance, i.e. the maximum of the particle likelihoods $p\left(\mathbf{y}_{n-1} \middle| \mathbf{x}_{n-1}^{(i)} \right)$

$$p\left(\mathbf{y}_n \middle| \mathbf{x}_g^{(k)} \right) > \tau_g * p\left(\mathbf{y}_{n-1} \middle| \mathbf{x}_{n-1}^{(i)} \right) \qquad (4.44)$$

These locations of good measurement match are used to bias the proposal distribution towards them. This is achieved by forming a sum of n_g Gaussians centred at the good grid locations and with covariance matrices $\mathbf{C_x}$. Hence the proposal distribution is given by:

$$q\left(\mathbf{x}_n \left|\mathbf{x}_{n-1}, \mathbf{y}_n\right.\right) =$$

$$= \beta_{RW} N\left(\mathbf{x}_n \left|\mathbf{x}_{n-1}, \mathbf{C_x}\right.\right) + \frac{1 - \beta_{RW}}{n_g} \sum_{k=1}^{n_g} N\left(\mathbf{x}_n \left|\mathbf{x}_g^{(k)}, \mathbf{C_x}\right.\right) \tag{4.45}$$

Note that in [Pérez *et al.* (2004)] the measurement cue employed for modifying the measurement-assisted proposal distribution is colour. This is by no means restrictive; any measurement cue can be used to find the good matching locations on the grid.

4.11.4 *Particle filter face tracker with colour-assisted proposal distribution*

To implement the grid search for good likelihood state-space points and the substitutions of some particles with Gaussian-perturbed versions of these good points, the face tracker of example 4.13 is modified as shown in example 4.14.

Example 4.14. Implementation of the SIR particle filter face tracker with the measurement-assisted proposal distribution discussed in section 4.11.3.

In this example the part of the code covering the proposal distribution sampling changes to turn the SIR particle filter face tracker of example 4.13 into a one with measurement-assisted proposal distribution is analysed.

The first step of the implementation is to draw the samples from the random walk component. The samples at the previous time step are originally x and they are replaced by the newly drawn ones in the usual way:

```
% Sample measurement-assisted proposal distribution for updated particles
% 1. Gaussian (random walk) component
x=x+sqrt(C)*randn(size(x,1),Np);
```

The second step is to draw samples from the measurement component. This is illustrated in Fig. 4.21. First of all the grid around the previous time instance state estimate xo is defined. The step for all the state dimensions is given in the vector **step** and the number of steps for all the state dimensions above and below the centre value in the vector **num**. The points along all dimensions are then stored in the cell array **g**, whose elements are vectors

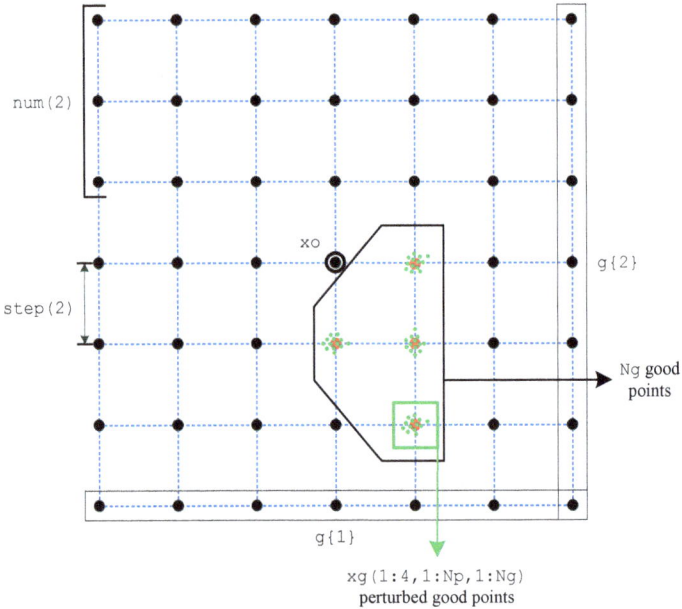

Fig. 4.21   Drawing samples from a measurement-assisted proposal distribution.

of variable length, one vector per state dimension. The implementation is
as follows:

```
% 2. Measurement component
% 2.a. Define grid
step=[1 1 .1 .1]'.*[xo(3) xo(4) xo(3) xo(4)]';
num=[2 2 1 1];
for i1=1:length(xo)
  g{i1}=round(xo(i1)-step(i1)*num(i1):step(i1):xo(i1)+step(i1)*num(i1));
end
```

The grid is then searched using a nested for-loop. In the loop each grid
point is mapped into a rectangular image region bounded by the box **bb**,
and a check is performed in order not to measure outside the frame I (i.e.
bb remains inside I). Subsequently the colour likelihood is measured and is
stored in **pyx**:

```
% 2.b. Measurement likelihood on grid
pyx=zeros(length(g{1}),length(g{2}),length(g{3}),length(g{4}));
for i1=1:length(g{1})
  for i2=1:length(g{2})
    for i3=1:length(g{3})
```

```
      for i4=1:length(g{4})
        bb=[g{1}(i1) g{2}(i2) g{3}(i3) g{4}(i4)];
        if bb(1)>=1 && bb(1)+bb(3)-1<=size(I,2)...
           && bb(2)>=1 && bb(2)+bb(4)-1<=size(I,1)
          Hface=histogram(I(bb(2):bb(2)+bb(4)-1,bb(1):bb(1)+bb(3)-1,:),Nh);
          pyx(i1,i2,i3,i4)=exp(-(1-sum(sqrt(Hface.*target.Hface)))/(2*s2c));
        end
      end
    end
  end
end
```

Following the grid search, the indices of the Ng points of good colour match are stored in ig and for each one of them Np Gaussian perturbations are stored in xg. The perturbations are obtained with a variance that is a fraction fgrid of the one used in the Gaussian random walk component. On average 1-bRW of these perturbations are going to replace the Gaussian random walk particles, randomly selected: each particle is retaining the Gaussian random walk sample with probability bRW, while it is changing to any of the good grid points with probability (1-bRW)/Ng. These steps are implemented as follows:

```
% 2.c. Find good grid points and draw particles around them
ig=find(pyx>pOMthr*max(pOM));
Ng=length(ig);
xg=randn(length(target.xo),Np,Ng);
if Ng>0
  % Perturb every good grid point Ng times
  [ig1 ig2 ig3 ig4]=ind2sub(size(pyx),ig);
  for ig=1:Ng
    xg(:,:,ig)=sqrt(C/fgrid)*xg(:,:,ig)+...
      repmat([g{1}(ig1(ig));g{2}(ig2(ig));g{3}(ig3(ig));g{4}(ig4(ig))],1,Np);
  end
  % Select a sample from x with probability bRW or from one of the xg
  % with probability (1-bRW)/Ng
  probs=cumsum([bRW (1-bRW)/Ng*ones(1,Ng)])';
  probs=[0;probs(1:end-1)];
  idx=rand(1,Np);
  for ig=1:Np
    t=find(repmat(idx(ig),length(probs),1)-probs>=0);
    idx(ig)=t(end);
    if idx(ig)>1
      x(:,ig)=xg(:,ig,idx(ig)-1);
    end
  end
end
```

∎

A very important result occurring with the use of the measurement-assisted proposal is that the performance of the tracker is no longer heavily dependent on the variance of the Gaussian random walk component, nor on the colour variance. This is very good, since the optimum values of these parameters change depending on the type of content being tracked (fast vs. slow motion for the former, constant vs. variable illumination for the latter). Furthermore, the effect of the fraction τ_g of the best match obtained at the previous time instance that needs to be exceeded for a grid point to be accepted is minimal.

The results of the measurement-assisted proposal distribution are shown in the last row of Table 4.6. Note that for the grid search in this case the best (and faster) version of the tracker does not search for independent width and height changes, i.e. the grid search is three-dimensional.

4.11.5   *Particle filter face tracker using face likelihood*

The second measurement cue used for face tracking in the face detection likelihood. Thus the likelihood is evaluated using (4.34). This is demonstrated in example 4.15.

Example 4.15. Evaluate the face likelihood of (4.34).

Assume that the Viola–Jones face detector discussed in section 4.8.1 returns many face bounding boxes stored in an array BB. These can be obtained by running the face detector from OpenCV, with the number of rectangles set to zero in order to disable merging of similar bounding boxes. The Np particles at the given instance are comprised of samples x. To evaluate the face likelihood pyx, the following loop is used:

```
pyx=zeros(1,Np);
if ~isempty(BB)
  for i1=1:Np
    wf=exp(-sum((repmat(x(3:4,i1)',size(BB,1),1)-BB(:,3:4)).^2,2)...
      ./prod(BB(:,3:4),2)/2/s2w)';
    D=1./(sqrt(sum((repmat((x(1:2,i1)-ceil(x(3:4,i1)/2))',size(BB,1),1)...
      -BB(:,1:2)).^2,2))/x(3,i1)+1);
    pyx(i1)=wf*D;
  end
end
```

The factor wf is the face size matching factor of (4.35), with s2w being the size matching variance σ_w^2, which is set to 0.04 for this example.

The factor D is the position matching factor of (4.34), with the distance difference exponent K set to unity. ∎

The particle filter using face likelihood is employed in the same sequence as the colour-based one. The performance is compared to the colour-based version in Table 4.7.

Table 4.7   SIR face tracker using colour or face likelihood.

Measurement	Position error (%)	Size error (%)	Spread (%)
Colour	15.5	27.6	32.5
Face	10.3	18.6	28.2

Although across the complete sequence the face likelihood results in a far better tracker, it is interesting to see how this changes per frame, averaged over 50 different runs. This is shown in Fig. 4.22 for the position and in Fig. 4.23 for the size errors. There are a few cases where the face detector does not give good detections because of the pose of the head. For those frames the performance of the face likelihood measurement is worse than for the colour one (in both position and size). This observation will be exploited in the multi-cue visual face trackers designed throughout Chapter 5.

Finally, some example frames with the particles on top are shown in Fig. 4.24. The particles are shown by dashed rectangles, blue for those corresponding to the Gaussian random walk component and green for those corresponding to the grid search. The state estimate is shown by a thick red rectangle. Both the colour cue (left column) and the face cue (right column) are shown for comparison. Note that in the first and fourth example frames the pose of the head is not favourable for the face detector, hence the framing of the face when the colour cue is used is better. In the rest of the example frames the pose is frontal and the framing of the face when using the face cue is better. This is reflected in the error comparison between the two particle filters in Figs. 4.22 and 4.23.

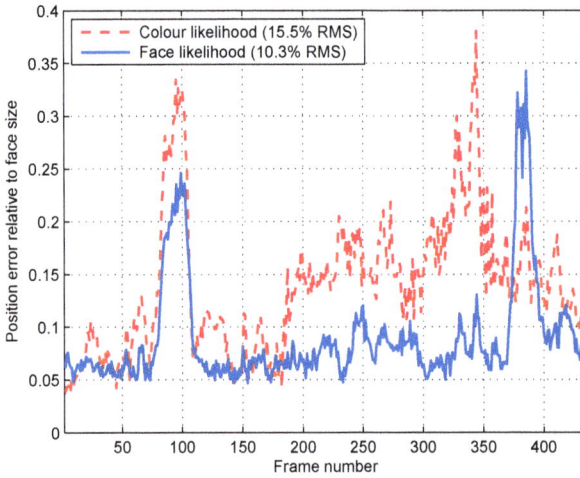

Fig. 4.22   Position error relative to the face size as a function of frame number, averaged over 50 runs of the particle filter for the colour and the face likelihood measurements. Face measurements lead to a better tracker with the exception of the error peak around frame 375.

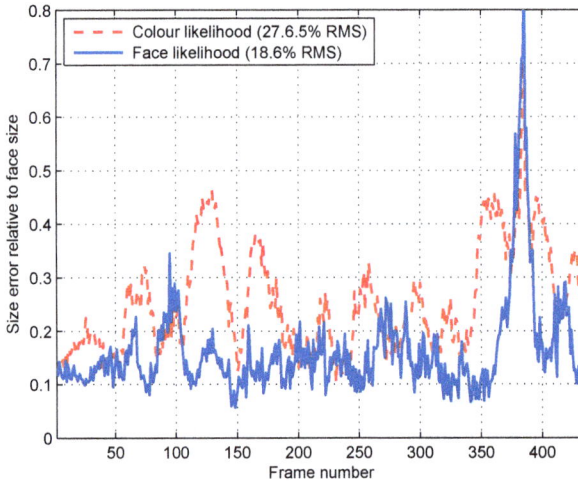

Fig. 4.23   Size error relative to the face size as a function of frame number, averaged over 50 runs of the particle filter for the colour and the face likelihood measurements. Face measurements lead to a better tracker with the exception of the error peak around frame 100.

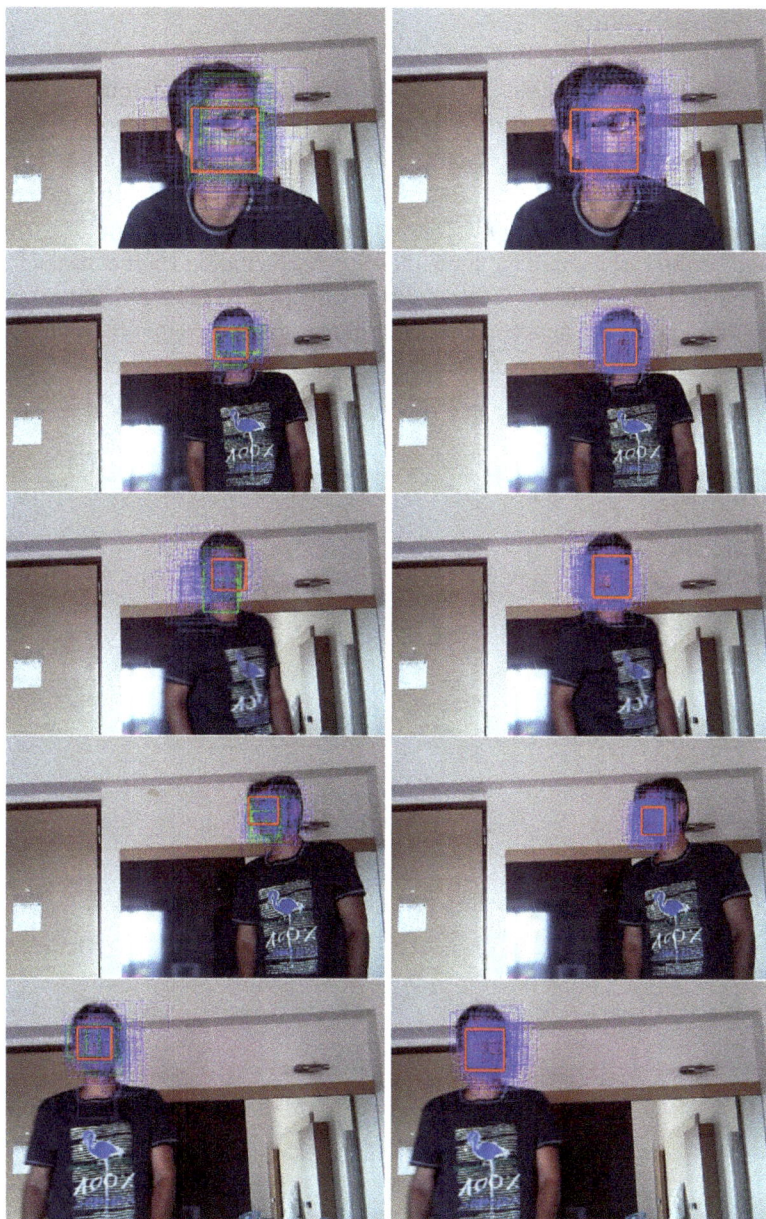

Fig. 4.24   Example frames showing the performance of the colour (left column) and the face (right column) particle filter.

4.12   Conclusions

Visual tracking is based on different possible measurements on moving images: colour, motion, outline and specific object (such as face) detection have been detailed in this chapter. Utilising any of these in a particle filter tracker requires the definition of a likelihood function. Such functions have been derived throughout this chapter. Tracking systems have then been described, that manage multiple targets, including their initialisation, tracking and termination. The chapter concludes with tracking examples, concentrating on the face tracker using different measurements, with different performance in different situations. These examples pave the way for tracking using multiple measurements, possibly across different modalities, in Chapter 5.

Chapter 5

Multimodal Tracking

5.1 Introduction

Tracking humans with a single measurement is not robust [Hua and Wu (2006); Veeraraghavan *et al.* (2006); Spengler and Schiele (2001)]. The audio and visual modalities both suffer from:

- Background clutter, e.g. similar shapes, colours or sounds.
- Occlusions or masking, e.g. other objects in the foreground or strong sounds.
- Temporal persistence, e.g. a person does not speak all the time, or his/her face is not visible from the camera as the pose changes.
- Appearance changes, e.g. illumination changes that change colour appearance and render colour models unsuitable.

Hence using a single measurement type can be beneficial under some conditions, but as these conditions change (and these changes can be a simple target movement relative to a camera) another measurement type can be of more use. A typical example of this has been demonstrated in Figs. 4.22, 4.23 and 4.24, where a face tracker is shown to operate robustly using face detection measurements, until the pose becomes non-frontal and then the colour measurement has superior performance.

Tracking humans by fusing multiple measurement types allows for one measurement type to assist in tracking when the conditions change and become less favourable for another measurement type. It also allows for more efficient handling of appearance changes by allowing for model retraining. When a measurement type results in a very high likelihood, then the model for another measurement type can be updated.

The different measurements can be performed on a single modality, e.g. some of the visual measurements analysed in Chapter 4, leading to multi-cue visual trackers. They can also be performed across the two different modalities considered in this book, leading to audio-visual trackers [Bernardin *et al.* (2008); Gatica-Perez *et al.* (2003); Zotkin *et al.* (2002); Beal *et al.* (2002)].

Fusing measurements can be tricky though, as the following questions need to be answered:

- Which measurement types to fuse and how?
- Can the tracking system afford the computational cost of multiple measurement types?
- Is there any guarantee that the fused tracker will not result in performance loss when the conditions are favourable for one of the measurement types, compared to the uni-modal version employing the favoured measurement type?

In the following sections three different ways of fusing different measurement types are considered:

- Likelihood combination within a single tracker (see secton 5.2).
- Combination of multiple trackers by decision fusion (see secton 5.3).
- Cascading trackers using different measurement types in the partitioned sampling framework (see secton 5.4).

The discussion will lead to different multi-cue and multimodal trackers.

5.2   Likelihood combination

The simplest way to fuse multiple measurement types, possibly across different modalities, is to generate a new likelihood by combining those of the different measurement sources. The block diagram of the fused tracker is shown in Fig. 5.1.

In the context of particle filtering it is possible to exploit the relation between the structure of the model and the information resulting from the different measurement sources. Assuming Ξ measurement sources, the measurement vector can be written as $\mathbf{y} = [\mathbf{y}^1, \ldots, \mathbf{y}^\Xi]$. Given the state, if measurements are considered conditionally independent the likelihood function can be factorised as [Pérez *et al.* (2004)]:

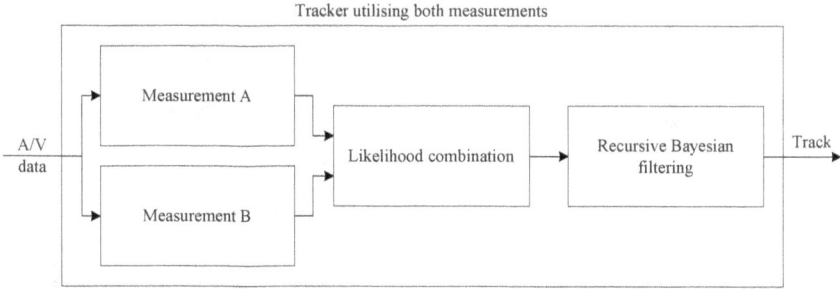

Fig. 5.1 Fusing different measurements (conducted around the state) by likelihood combination within a single tracker.

$$p(\mathbf{y}|\mathbf{x}) = \prod_{\xi=1}^{\Xi} p(\mathbf{y}^{\xi}|\mathbf{x}) \qquad (5.1)$$

Thus the generic particle filtering framework can be used with the weight update involving Ξ likelihood calculations. This in fact represents the likelihood calculations as used in section 3.4 for the case of audio tracking where TDE and BF calculations were performed for each of the Ξ particles and combined as in (5.1).

5.2.1 *Multiplying face and colour likelihoods*

As an example of likelihood combination fusion as in (5.1) and Fig. 5.1, the face and colour likelihoods are calculated for the frame in Fig. 5.2. Observe that the colour likelihood peaks at several background objects, while the face likelihood misses the profile face entirely. Their product, also depicted in Fig. 5.2 offers two clear peaks at the two frontal faces and a rather smaller one at the profile face.

Employing such a likelihood product in a multi-cue visual face tracker utilising the face and colour likelihoods should thus attenuate the effect of background clutter and pose variations. To facilitate the likelihood combination by multiplication, the face likelihood is made less selective than the uni-modal face tracker of section 4.11.5. With regards to the size matching factor of (4.35), the variance is made larger than the uni-modal case, by setting $\sigma_w^2 = 0.09$. With regards to the position matching factor in the denominator of (4.34), the exponent of the distance difference is set to $K = 1/2$.

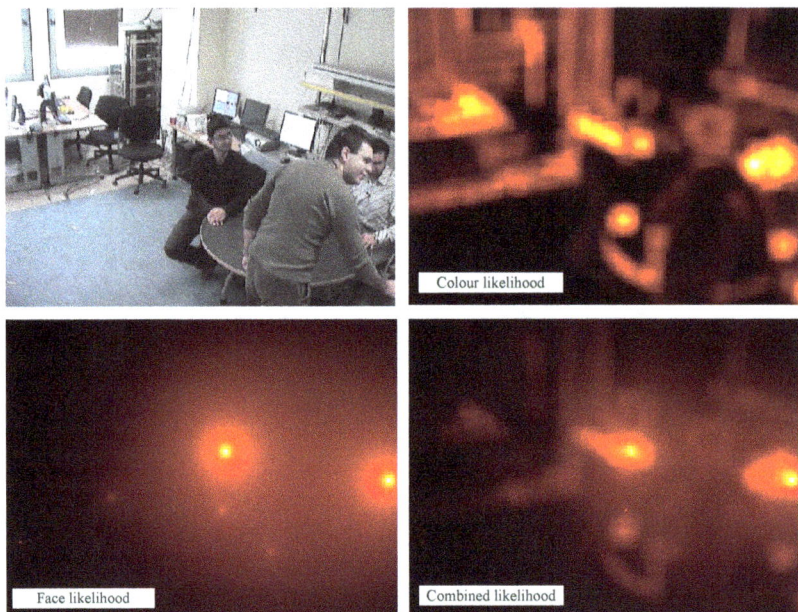

Fig. 5.2   Likelihood combination by multiplying colour and face detection likelihoods.

Table 5.1   Comparison of multi-cue visual face trackers with different fusion strategies. For partitioned sampling, the cue used first for position update is indicated.

Fusion strategy	Position error (%)	Size error (%)	Spread (%)
Colour only	15.5	27.6	32.5
Face only	10.3	18.6	28.2
Likelihood combination	10.3	18.3	27.6
Output combination	10.8	16.2	35.7
Partitioned sampling, colour→pos	10.0	17.2	24.2
Partitioned sampling, face→pos	20.2	28.3	28.8

Running the likelihood combination multi-cue face tracker on the same sequence as the uni-modal face trackers in Chapter 4 results in the performance in the third row of Table 5.1 and the per frame estimation errors in Fig. 5.3 for position and Fig. 5.4 for size.

Both the results averaged throughout the sequence and the estimation errors per frame indicate that fusion by likelihood multiplication improves slightly in position estimation over the best measurement cue, the face detections. Only size estimation improves notably.

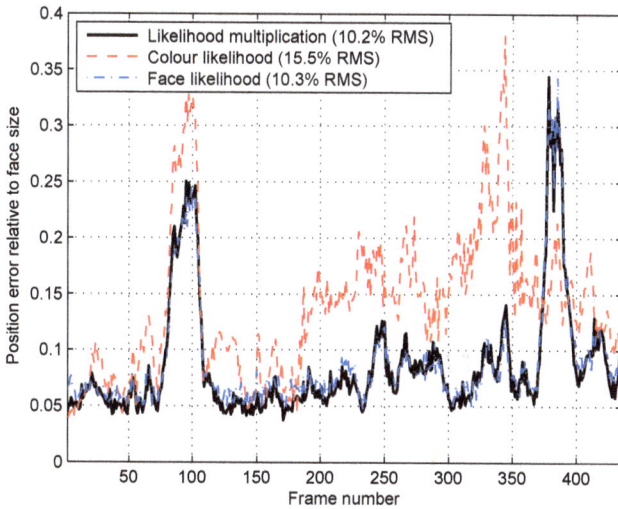

Fig. 5.3 Likelihood combination face tracker: position error relative to the face size as a function of frame number, averaged over 50 runs of the particle filters for the single measurement (colour and face likelihoods) and their combination by multiplication.

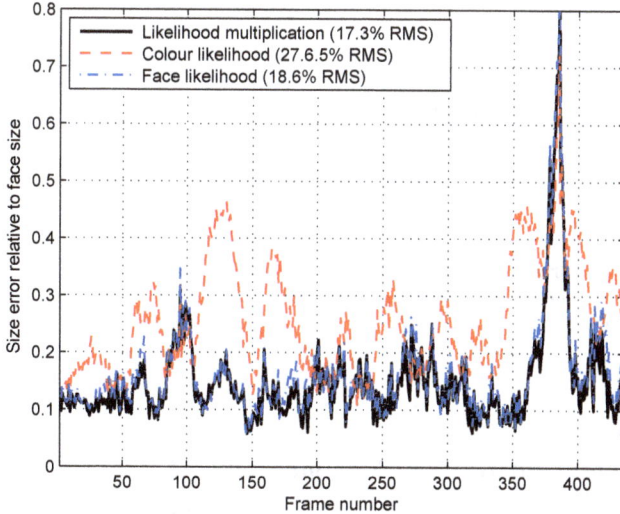

Fig. 5.4 Likelihood combination face tracker: size error relative to the face size as a function of frame number, averaged over 50 runs of the particle filters for the single measurement (colour and face likelihoods) and their combination by multiplication.

5.3   Tracker output combination

The various measurement types can also be considered independently, by building trackers utilising them and subsequently combining the tracker outputs [Badrinarayanan *et al.* (2007); Babu *et al.* (2007); Leichter *et al.* (2006)]. This is shown in the block diagram of Fig. 5.5. The decision fusion module not only combines the decisions, but also possibly controls the states of the individual trackers.

The scheme of Fig. 5.5 is not as straightforward in the case of multiple trackers handling different targets. Assuming K targets and Ξ measurement types, there is a total of $K \times \Xi$ tracker outputs that need to be associated in K groups of Ξ elements each, containing at most one track from every measurement type. Any group outliers can be discarded as measurement types that lead to bad tracking.

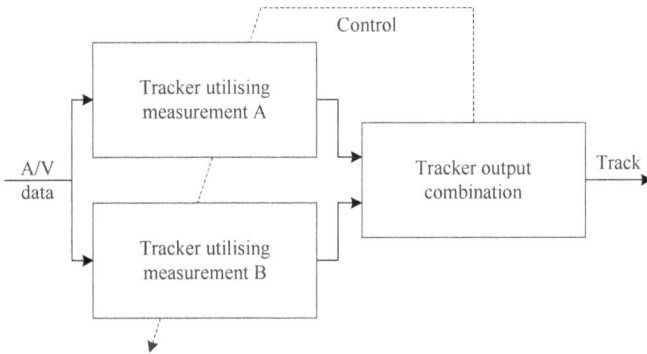

Fig. 5.5   Fusing trackers employing different measurements by combining the outputs of the trackers. The combination logic can potentially lead to a feedback controlling the state of the trackers.

5.3.1   *Deterministic fusion of independent trackers*

There are two ways to combine the outputs of the trackers. The deterministic combination is based on empirical rules on the involved likelihoods (see section 5.3.1.1) or on the geometry of the position estimations (see section 5.3.1.2). Stochastic combination is achieved using the decentralised Kalman filter discussed in section 5.3.2.

5.3.1.1 *Combining independent face trackers employing different visual measurements*

Returning to the combination of the face and the colour likelihood visual face trackers, the output combination scheme employs the maximum face likelihood in each time step. If this likelihood is above a high threshold, then the face detections accurately match one of the particles. This is an indication that the face measurement cue can be trusted as a standalone, hence the decision fusion returns the output of the face likelihood tracker. If it is below a low threshold, then the face measurement should be discarded and the decision fusion returns the output of the colour likelihood tracker. For face likelihoods in between the two thresholds the two outputs are averaged. The decision is also fed back to the trackers; the particles of any of the individual trackers are reset to those of the others if the face likelihood is too high or too low respectively.

Running the independent tracker combination multi-cue face tracker on the same sequence as the likelihood combination one, results in the performance in the fourth row of Table 5.1 and the per frame estimation errors in Fig. 5.6 for position and Fig. 5.7 for size.

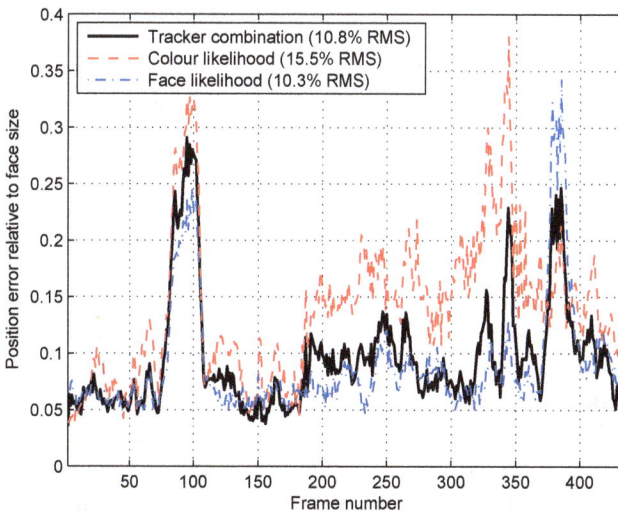

Fig. 5.6   Tracker output combination: position error relative to the face size as a function of frame number, averaged over 50 runs of the particle filters for the single measurement (colour and face likelihoods) and their post-decision fusion.

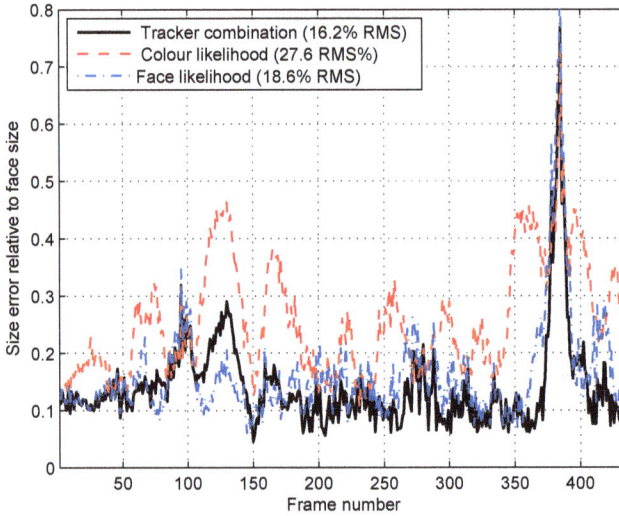

Fig. 5.7   Tracker output combination: size error relative to the face size as a function of frame number, averaged over 50 runs of the particle filters for the single measurement (colour and face likelihoods) and their post-decision fusion.

Although position estimation suffers slightly compared to the uni-modal tracker utilising face likelihood, size estimation improves drastically throughout the sequence. As for individual frames, the position error peak around frame 380 resulting from difficult to detect face pose is attenuated. On the other hand, the good position estimation of the face measurement in favourable frames is somewhat lost as the tracker output combination performance is between that of the face and the colour tracker.

5.3.1.2   *Combining independent audio and visual trackers in 3D space*

The goal in audio-visual tracking is to track the active speaker. Any visual system is tracking multiple people, of which only one is speaking at any time. This person is tracked by an audio system. The audio-visual system associates the audio track with one of the visual tracks and subsequently combines the two providing better state estimation.

The audio and visual face tracking systems described in Chapters 3 and 4 respectively give location estimates with some uncertainty [Talantzis *et al.* (2008); Pnevmatikakis and Talantzis (2010)]. The audio tracker utilises three microphones arranged in two pairs, yielding the position of the speaker on the floor, with height uncertainty. The visual face tracker yields faces on the image plane. With reference to Fig. 5.8, the visual position on the image plane (due to the depth uncertainty of the camera) corresponds to any point along the red line connecting the origin of the camera coordinate system $[x_v, y_v, z_v]$ with the depth-normalised coordinates $\mathbf{v}_n$ from the visual track (see Fig. 4.1 and 3D position face validation in section 4.8.2). The image plane coordinates are transformed to $\mathbf{v}_n$ using the intrinsic camera parameters [Zhang (2000)]. Similarly, the audio position (with the height uncertainty of the audio tracker) corresponds to any point along the green line connecting the origin of the audio coordinate system $[x_a, y_a, z_a]$ with the height-normalised coordinates $\mathbf{a}_n$ from the audio track.

Assuming that the two independent audio and visual trackers yield one output each, the two tracks are combined utilising the intersection of the two uncertainty lines, effectively eliminating the location uncertainty. Ideally the two lines intersect, but in practice audio and visual tracking errors and the different targets (mouth for the audio and centre of vaguely frontal face for the visual tracker) result in no intersection. We instead employ a least squares solution to find the point of minimum distance from both lines, i.e. the centre $[x_{av}, y_{av}, z_{av}]$ of the minimum length segment connecting the uncertainty lines. This minimum length is $2e_{av}$, and used as a measure of the quality of match of the audio location with the visual one.

To formulate the problem, we need to relate both the audio and the visual coordinate systems to the world coordinate system $[x, y, z]$. This is done by finding the respective translation vectors $\mathbf{T}_a$ and $\mathbf{T}_v$ (green and red dashed lines) as well as the rotation matrices $\mathbf{R}_a$ and $\mathbf{R}_v$. For the visual coordinate systems, these are the extrinsic parameters of the camera [Zhang (2000)]. For the audio system, the translation vector is simply the location reported by the audio tracker on the floor-plan, while the rotation matrix is the identity one, since the orientation of the system does not change with respect to the world one. The normalised audio coordinates are then the z-axis unity vector. The least squares solution is then obtained by solving:

$$\begin{bmatrix} \mathbf{I}_3 & -\mathbf{R}_v\mathbf{v}_n & \mathbf{0}_3 \\ \mathbf{I}_3 & \mathbf{0}_3 & -\mathbf{R}_a\mathbf{a}_n \end{bmatrix} \cdot \mathbf{x}_{av} = \begin{bmatrix} \mathbf{T}_v \\ \mathbf{T}_a \end{bmatrix} \tag{5.2}$$

for $\mathbf{x}_{av}$, where $\mathbf{I}_3$ is the 3×3 identity matrix and $\mathbf{0}_3$ is the 3×1 zero vector. The system is readily solved using the pseudo-inverse of the left-hand matrix.

The more usual case though is to have multiple video face tracks and a single audio track. All face tracks are then matched to the audio one and the fusion is reported for the face track resulting in the minimum e_{av}. It is also possible to place an upper bound to the accepted e_{av}: if the distance between the best matching face track and the audio track is larger than this upper bound then the audio source is regarded as something not coming from any of the people present in the scene.

An example of the approach is shown in Fig. 5.9. The visual face tracker reports three faces while the audio tracker reports a sound source that can be associated with the furthest right face. This results in a refinement of the audio/visual track of the main speaker.

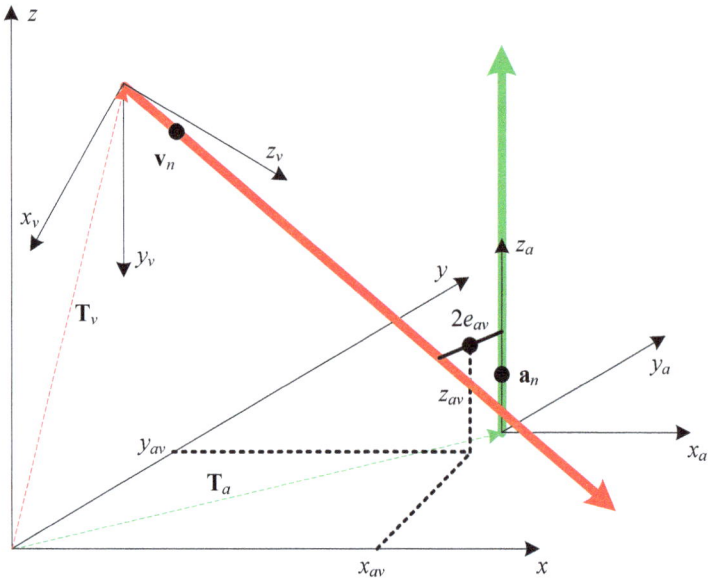

Fig. 5.8   Normalised audio ($\mathbf{a}_n$) and visual ($\mathbf{v}_n$) estimations with their uncertainties (green and red thick lines) on the respective coordinate systems ([x_a, y_a, z_a] and [x_v, y_v, z_v]), offset and rotated with respect to the world coordinate system [x, y, z]. The audio-visual fusion is the centre [x_{av}, y_{av}, z_{av}] of the minimum length ($2e_{av}$) segment connecting the uncertainty lines.

5.3.2   *Kalman fusion of independent audio and visual trackers*

The location estimates provided by 3D audio and video trackers can be recursively combined by the use of a decentralised Kalman filter [Brandstein

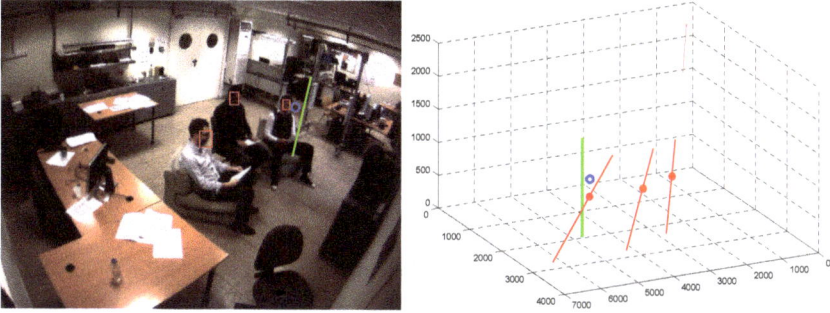

Fig. 5.9 Example of fusing the independent audio and video tracks: representation on the image plane and the 3D space.

and Ward (2001)] as described in [Katsarakis *et al.* (2008)]. In this section we attempt a brief overview of the architecture presented there. The overall fusion system can be seen in Fig. 5.10. It comprises two local Kalman filters and a two-input global one. The local Kalman filters operate on the outputs of the 3D audio and video trackers, yielding a set of weighted observation vectors for each modality at time instant n:

$$\mathbf{q}_\phi = \mathbf{P}_\phi^{-1}[n+1|n+1]\mathbf{y}_\phi[n+1|n+1] - \mathbf{P}_\phi^{-1}[n+1|n]\mathbf{y}_\phi[n+1|n] \quad (5.3)$$

where $\phi = audio, video$ refers to each of the modalities. In the above equation $\mathbf{y}_\phi[n|n]$ is the previous standalone modality state, $\mathbf{y}_\phi[n+1|n]$ is the state updated by the dynamic model and $\mathbf{y}_\phi[n+1|n+1]$ is the next state, after the measurement update. Also, $\mathbf{P}_\phi[n+1|n+1]$ is the a posteriori error covariance matrix after measurement update of $\mathbf{P}_\phi[n+1|n]$ using:

$$\mathbf{P}_\phi^{-1}[n+1|n+1] = \mathbf{P}_\phi^{-1}[n+1|n] + \mathbf{H}_\phi^T \boldsymbol{\Lambda}_\phi^{-1} \mathbf{H}_\phi \quad (5.4)$$

with:

$$\mathbf{H} = \begin{bmatrix} 1 & 0 & 0 & 0 \\ 0 & 0 & 1 & 0 \end{bmatrix} \quad (5.5)$$

The $\boldsymbol{\Lambda}_\phi^{-1}$ diagonal matrices for the audio and video measurement uncertainties effectively adjust the trust we put in any of the two modalities. The global KF provides fused weighted observation vectors given as:

$$\mathbf{q} = \mathbf{P}^{-1}[n+1|n+1]\mathbf{y}[n+1|n+1] - \mathbf{P}^{-1}[n+1|n]\mathbf{y}[n+1|n] \quad (5.6)$$

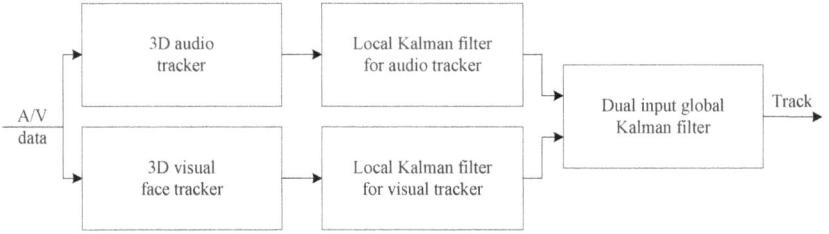

Fig. 5.10 Block diagram of the decentralised Kalman filter used for combining the outputs of the 3D audio and video trackers.

where $\mathbf{P}$ is the global a posteriori error covariance matrix. The above can be solved for the fused state $\mathbf{y}[n+1|n+1]$ to give:

$$\mathbf{y}[n+1|n+1] = \mathbf{P}[n+1|n+1](\mathbf{P}^{-1}[n+1|n]\mathbf{y}[n+1|n] + \mathbf{q}[n+1]) \quad (5.7)$$

where $\mathbf{q}[n+1] = \mathbf{q}_{audio}[n+1] + \mathbf{q}_{video}[n+1]$ while the global a posteriori error covariance matrix is updated according to:

$$\begin{aligned}
\mathbf{P}^{-1}[n+1|n+1] &= \mathbf{P}^{-1}[n+1|n] \\
&+ (\mathbf{P}^{-1}_{audio}[n+1|n+1] - \mathbf{P}^{-1}_{audio}[n+1|n]) \\
&+ (\mathbf{P}^{-1}_{video}[n+1|n+1] - \mathbf{P}^{-1}_{video}[n+1|n])
\end{aligned} \quad (5.8)$$

5.4 Partitioned sampling

The reasoning behind likelihood combination in (5.1) can be extended further by assuming we can split the particle filtering object model into Ξ successive intermediary steps i.e. assume that the state can be partitioned into Ξ groups of variables (hence it needs to be at least Ξ-dimensional) and the corresponding Ξ evolution models are independent.

If we now make the approximation that the likelihood for the ξth measurement modality $p(\mathbf{y}^\xi|\mathbf{x})$ can be incorporated after applying the ξth state evolution model $p_\xi(\mathbf{x}^\xi|\mathbf{x}^{\xi-1})$, we can set up a recursion to compute the new target distribution as follows [Pérez *et al.* (2004); MacCormick and Blake (2000)]:

$$\zeta^\xi(\mathbf{x}^\xi) \propto \int w_\xi(\mathbf{x}^\xi, \mathbf{x}^{\xi-1}) q_\xi(\mathbf{x}^\xi|\mathbf{x}^{\xi-1}, \mathbf{y}^\xi) \times \zeta^{\xi-1}(\mathbf{x}^{\xi-1}) d\mathbf{x}^{\xi-1}, \xi = 1, \dots \Xi$$

$$(5.9)$$

with:

$$w_\xi(\mathbf{x}^\xi, \mathbf{x}^{\xi-1}) = \frac{p(\mathbf{y}^\xi|\mathbf{x}^\xi)p_\xi(\mathbf{x}^\xi|\mathbf{x}^{\xi-1})}{q_\xi(\mathbf{x}^\xi|\mathbf{x}^{\xi-1}, \mathbf{y}^\xi)} \qquad (5.10)$$

In the above ζ^0 and ζ^Ξ are the previous and the new filtering distributions respectively. The recursion can be approximated with a layered sampling strategy where during the ξth stage new samples are simulated from a Monte–Carlo approximation of $q_\xi(\mathbf{x}^\xi|\mathbf{x}^{\xi-1}, \mathbf{y}^\xi)\zeta^{\xi-1}(\mathbf{x}^{\xi-1})$ with an associated importance weight proportional to $w_\xi(\mathbf{x}^\xi, \mathbf{x}^{\xi-1})$.

The process is illustrated in Fig. 5.11. It is not straightforward what the partition of the state-space can be, nor which of the available measurement cues should be updating which partition. An example is given in the next section.

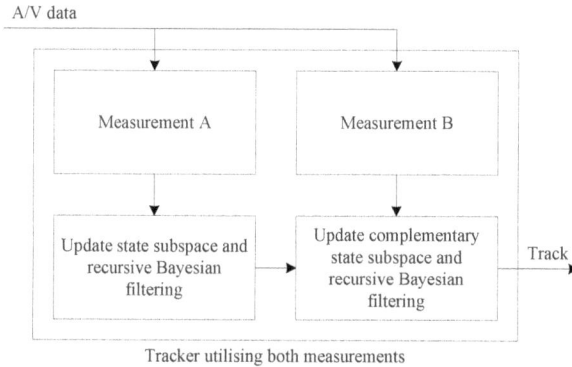

Fig. 5.11 Fusing different measurements by cascading particle filter trackers according to the partitioned sampling framework. Each measurement updates a subspace of the state within a particle filter tracker. The updated state subvector has more state variables appended and the process is repeated utilising a different measurement.

5.4.1 *Joint multi-cue face tracker employing partitioned sampling*

The final multi-cue face tracker considered in this Chapter uses again the colour and face likelihoods, this time in a partitioned sampling framework.

The two size dimensions are not independent, hence they cannot be partitioned. The two position dimensions are themselves independent and are also independent from the size ones. Hence there are several ways to partition the state-space. We select to partition first the position dimensions, update them, and then append the size dimensions.

Two versions are generated, depending on which measurement is applied first. The first version follows the intuitive approach of [Pérez *et al.* (2004)] to first apply the more coarse measurement that does not offer precise localisation in position and size, i.e. the colour, which is in fact always available for measurement. This is followed by the fine localisation offered by the face detection measurement, which is not guaranteed to return any detection under adverse pose conditions. The second version applies the two measurements in reverse order. The performance of both versions is averaged over 50 runs in the final two rows of Table 5.1. From the results, evidently the second version is discarded as the results are worse than for both the mono-modal trackers!

The estimation errors per frame are given in Fig. 5.12 for position and Fig. 5.13 for size. In both estimations the partitioned sampling fusion is never worse than for the uni-modal tracker using the face measurement. On the other hand, both the major position estimation error peak around frame 380 and some other minor ones are alleviated, resulting in better performance for both position and size estimation.

In Fig. 5.14 an example face track is given for the estimation error peak of frame 385. Partitioned sampling fusion is framing the profile face better than any of the two uni-modal trackers do.

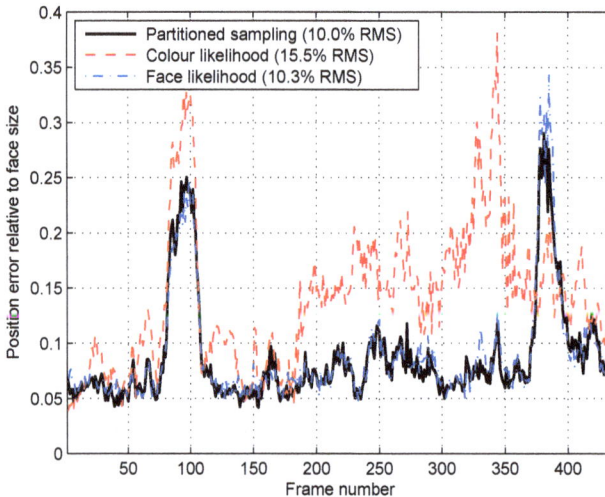

Fig. 5.12   Partitioned sampling face tracker: position error relative to the face size as a function of frame number, averaged over 50 runs of the particle filters for the single measurement (colour and face likelihoods) and their partitioned sampling fusion.

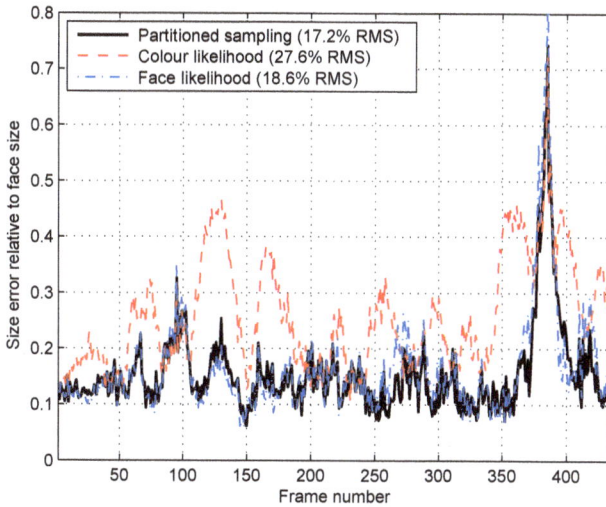

Fig. 5.13 Partitioned sampling face tracker: size error relative to the face size as a function of frame number, averaged over 50 runs of the particle filters for the single measurement (colour and face likelihoods) and their partitioned sampling fusion.

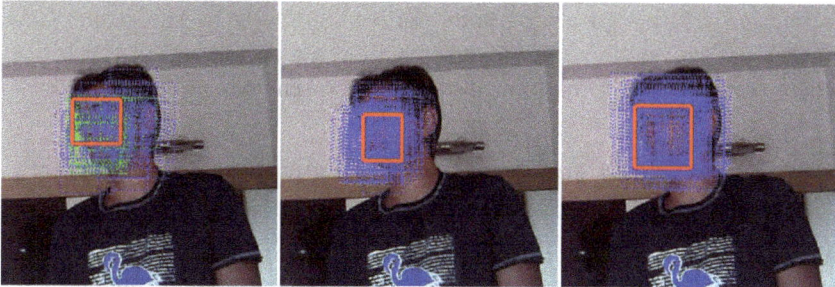

Fig. 5.14 Face track for frame 385 with the two uni-modal particle filter face tracker using colour (left) or face (middle), as well as the partitioned sampling fusion (right). The best framing by far is the fused tracker.

To compare all three visual multi-cue face trackers, the per frame estimation errors for likelihood combination, tracker combination and partitioned sampling are collected in Fig. 5.15 for position and Fig. 5.16 for size. Partition sampling and likelihood combination perform best in terms of position estimation, with the latter not managing very well at the peak around frame 380, where the face is in profile. Tracker output combination works best there, but it has other error peaks, overall ranking third.

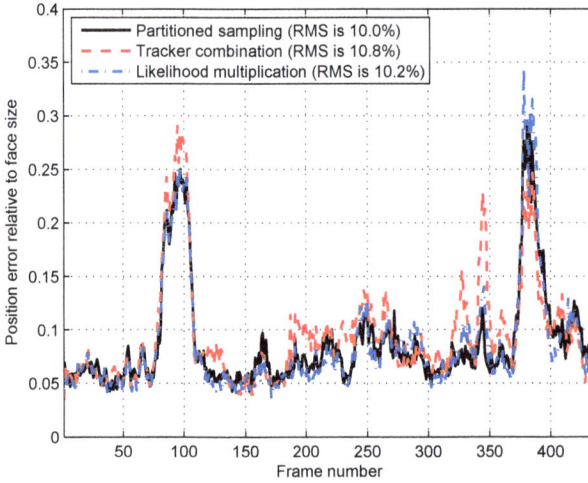

Fig. 5.15 Position error relative to the face size as a function of frame number, averaged over 50 runs of the particle filters for the three multi-cue (colour and face likelihoods) visual face trackers discussed in sections 5.2.1, 5.3.1.1 and 5.4.1.

On the other hand, tracker output combination is the clear winner for size estimation. Although all methods perform similarly at the error peaks, tracker output combination is somewhat better on almost all the frames where performance is good for all three fusion schemes.

5.4.2 *Joint audio and visual tracker employing partitioned sampling*

Assume a tracking system comprising a single camera and a microphone pair, with the x_c axis of the camera coordinate system coinciding with the line connecting the two microphones. This is depicted in Fig. 5.17. A direction of arrival estimate from the two microphones (represented by a vector $\mathbf{e}_{DOA}$) is then measured as the angle from the principal ray (axis z_c) of the camera. Due to the height uncertainty of the audio source estimate, this corresponds to any point on the plane defined by the y_c axis and the $\mathbf{e}_{DOA}$ vector.

The audio estimation-plane intersects the image-plane at a line perpendicular to the x_p axis and parallel to the y_p one. This corresponds to a horizontal position in the image-plane. The mapping is to the exact vertical line in the image, is non-linear and is governed by the intrinsic parameters of the camera.

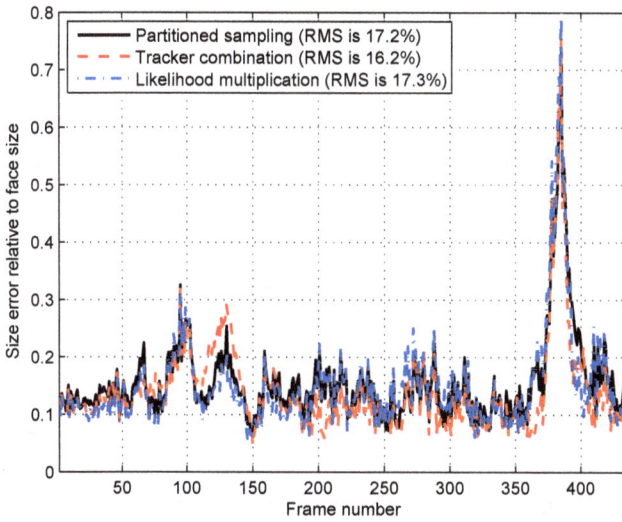

Fig. 5.16   Size error relative to the face size as a function of frame number, averaged over 50 runs of the particle filters for the three multi-cue (colour and face likelihoods) visual face trackers discussed in sections 5.2.1, 5.3.1.1 and 5.4.1.

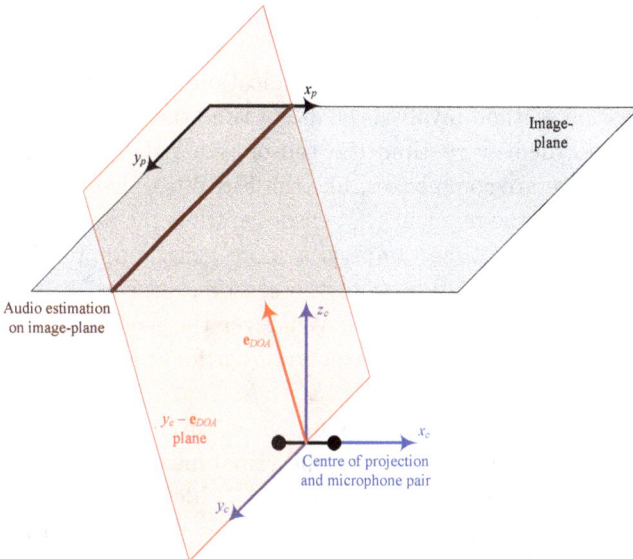

Fig. 5.17   Corresponding the direction of arrival to a line on the image-plane for the single camera and microphone pair setup of the joint audio and visual tracker employing partitioned sampling.

Using this hardware an audio-visual face tracking system is built employing partitioned sampling. We follow the same coarse-to-fine strategy as in the multi-cue visual face tracker. The state is partitioned in the horizontal position (a non-linear function of the direction of arrival) and the vertical position together with the two size variables. Horizontal position is updated using the audio modality. The other three state variables are then appended and the second stage utilises face detection likelihood.

5.5 Sensor synchronisation

In trackers that utilise multiple sensors, either by combining likelihoods or utilising partitioned sampling (i.e. there are no independent trackers operating on each measurement stream), care must be taken to synchronise the audio and visual measurements. This is the case when multiple cameras and/or microphone arrays are involved. Synchronisation is not an easy task as it potentially involves heterogeneous sensors connected to multiple computers.

Synchronisation of the start time is achieved by beginning the two measurements simultaneously. This is easy for sensors located at the same computer, but requires network communication if they are not. Another synchronisation method involves the use of timestamps. Usually the timestamps are recorded every time the sensor sends the information to the computer. There are some buses, like the FireWire, that provide the timing information at the moment the sensor captures the data, which is more accurate. The use of timestamps is a more generic approach, as it also accounts for lost information from the sensors, such as dropped frames, which is not an uncommon event. Whatever the method, the computers involved must have their clocks synchronised, for example using Network Time Protocol.

In the case of sensors that sample their environment at different rates, these also need to be synchronised. The visual measurements are obtained at a sub-multiple k_v of the frame rate $f_s^{(v)}$. Ideally $k_v = 1$ and the visual tracker is fast enough to process all frames. For audio measurements, the rate is typically faster, depending on the sampling rate $f_s^{(a)}$ and the number of samples k_a comprising the audio frame for direction of arrival measurement. To synchronise the two measurement processes, the larger processing interval $k_v/f_s^{(v)}$ is used as reference, and all the n_a audio measurements falling in this interval are averaged. It is:

$$n_a = \lfloor \frac{f_s^{(a)} k_v}{f_s^{(v)} k_a} \rfloor \tag{5.11}$$

where $\lfloor x \rfloor$ is the largest integer smaller than or equal to x.

5.6 Conclusions

Tracking humans benefits a lot from employing measurements obtained from multiple sensors, such as multiple microphone pairs and multiple cameras. For example, only with multiple sensors can we have 3D tracking. Furthermore, multiple measurement cues can be even more important, since there are no measurement cues suitable for all visual and audio target appearances. In this Chapter three different methods for combining different measurement cues (either all visual or multimodal) have been considered, all leading to performance improvements over any of the uni-modal trackers. Out of the three fusion schemes, a carefully designed partitioned sampling tracker gives the best performance.

Chapter 6

Applications

Even when examined in the context of rich audio-visual datasets, person tracking can remain an abstract concept. Its applications and feasibility can be better realised when considering example scenarios. In this chapter we perform a review of the most popular application areas of audio-visual person tracking. Along with the current status of each scenario, we also examine future directions and the factors that limit each of the systems.

In all of the examined scenarios/applications tracking the locations of people is a key prerequisite in achieving so-called context-awareness [Crowley (2003)]. A context-aware system senses the environment and reacts to it. In realistic context-aware applications, the person tracking task has to be carried out in a non-obtrusive fashion. Person tracking could be performed with devices carried by people but the algorithms and systems discussed throughout this book concentrate on far-field and unconstrained audio-visual versions that use sensors which are optimally hidden from the people being tracked.

The purpose of this chapter is not to introduce additional algorithmic frameworks but rather to put person tracking in a more practical and commercial perspective. Nevertheless, all of the theoretical tools presented in previous chapters are directly applicable in any of the applications presented in the forthcoming sections.

6.1 Smart Spaces

The emerging Ubiquitous Computing (UC) paradigm aims to exploit casually accessible sensors in a domestic or professional environment in order to transparently provide computing services, regardless of time and location of the user. A core characteristic of a UC environment is context sensitivity,

175

which refers to the ability of ubiquitous devices and systems to react to their environment and adapt their behaviour accordingly [Dey (2001b)]. A key prerequisite to this vision is the ability to detect the location of people and track their trajectories. The variable acoustic and visual conditions encountered in the environments humans live in, are addressed by multimodal trackers.

Probably the most characteristic application of the UC paradigm is the concept of Smart Spaces [Pnevmatikakis *et al.* (2009)]. This includes homes and offices that limit in size the idea of ubiquity but have remained for years the environments with most attention in the research community. Researchers working on the design of Smart Spaces envision buildings where the location and identity of people are constantly available in order to enable the employment of related services. The information is derived implicitly, without requiring end-users to provide explicit input. We review two characteristic scenarios namely Smart Homes and Smart Offices.

6.1.1 *The Smart Home*

The vision of intelligent houses that facilitate the life of their inhabitants has been a popular idea for decades. Recent efforts were further assisted by the construction of houses that embed sensors and connectivity in the same manner they include electricity. Additionally, due to the steer of focus towards our ageing society the creation of Smart Houses found a new role i.e. the support of elderly in their domestic environment [AAL (2010)]. Houses are now designed so that they assist people with reduced physical functions, with commercial solutions (including monitoring, emergency alarms and urgent doctor communications) being offered [Build-IT (2010)]. Research is being carried out to take these services one step further, to help senior citizens resolve the social isolation and the cognitive decline they face [HERMES (2010)].

As attractive as this scenario is, the software for an automated home must be tailored to a particular home and family, and updated as the family's lifestyle changes. Tackling the programming task is far beyond the capabilities and interest of typical home inhabitants. Indeed, even rudimentary forms of regulation, such as operating a timer thermostat, are inordinately difficult for people. The alternative of hiring professional technicians to update programs as necessary is used in some commercial systems, but is costly and inconvenient. Partly due to these difficulties, home automation has only limited acceptability. Due to the aforementioned facts researchers today attempt to "hide" the computers and their programming in the background [CHIL (2007)].

A central requirement to the above approach is the employment of a person tracking system that detects the location of multiple people and tracks them as they move inside the house. Keeping in line with the previous chapters we concentrate on the case where the person tracking task is performed using cameras and microphones. A number of Smart Homes have now been developed. The most interesting efforts are still non-commercial and reside in universities and research labs.

In the Aware Home research project supported by the Georgia Institute of Technology [Kientz *et al.* (2008)], researchers are working in a real three-floor house that functions as a living laboratory for the development of technologies that can be applied to domestic technologies. Focusing mainly on the scenario of elderly support, they have equipped the house with audio-visual sensors and a set of tracking and sensing technologies. The EasyLiving project has been one of the earliest efforts from Microsoft [Brumitt *et al.* (2000)] to integrate some level of intelligence into small rooms. Researchers are working on both microphone arrays for tracking active speakers and video tracking using stereoscopic cameras.

Another ambitious effort is the PlaceLab project of MIT's House_n initiative [Larson and Topping (2003)]. Microphones and cameras are hidden in cabinets in order to minimise the sense of obtrusiveness in the house. The house is inhabited by students that create a huge database of audio-visual data through their everyday activities. The data is subsequently used by researchers for a variety of technology development including person tracking. Another similar system is being developed at the University of Colorado where neural networks are constantly trained by a variety of sensors and components in order to match the needs and habits of inhabitants.

Several other systems with the same scope are designed in universities and company labs. The applications of their tracking-related audio-visual components can be summarised as follows:

- In the simplest scenario, the Smart House tracks its inhabitants in order to control lighting and electrical appliances. Only lights and devices in rooms with people in them remain switched on in order to implement to a great extent the environmentally-friendly aspect of modern buildings.
- A similar approach is followed in so-called multi-room entertainment. Someone detected and tracked in the house decides to listen

to some music or access some form of video feed. Given that the house can distinguish between different tracked people the selected media move with the user as he or she moves from room to room. The Smart House stops play in the previous room and transfers media to the available devices in the new one.

- In collaboration with other components of a Smart House, a person tracking system can become a key component of a security/monitoring system for detecting intruders. Using components for face recognition and video recordings the Smart House would be able to recognise and record unauthorised access and possibly trigger alarms. A multi modal version of the tracker can ensure these characteristics are based on acoustic or visual events.

- In scenarios where the house inhabitants are elderly or disabled, a person tracker can detect accidents, people falling or verbally asking for help. Again, suitable alarms can immediately be generated. An example of a 3D body tracker providing location and body posture metadata that can be used for accident detection is shown in Fig. 6.1.

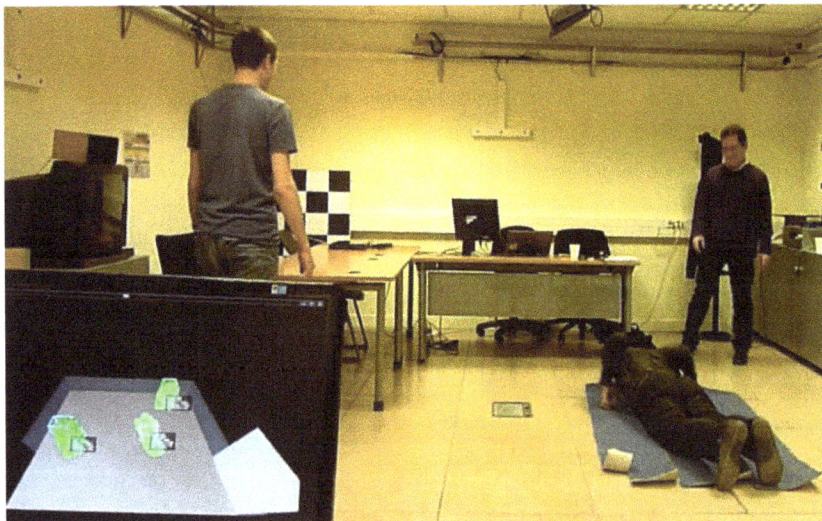

Fig. 6.1   3D tracker providing the locations and body postures (standing, walking, sitting, fallen) of all the people in the monitored space. The system combines motion evidence from several cameras into 3D human body abstractions [Andersen *et al.* (2010)].

- A central element to the vision of a Smart House is an Automatic Speech Recognition (ASR) engine. People can dictate commands that are recorded and recognised. Modern ASR systems only perform well with close-talking microphones, a fact that can potentially limit the Smart House experience. As demonstrated a person tracking system can estimate with great accuracy the position of the active speaker which can serve as an enabling prerequisite for far-field ASR. Technologies such as beamforming require this estimate in order to improve the quality of far-field speech signals and thus improve the accuracy of the corresponding ASR process.
- Person tracking also serves as a key component for training Smart Home models that monitor the habits of the inhabitants and tailor their behavior accordingly. Assuming for example that a cleaner appears periodically in the house and exhibits a specific tracking behaviour, the system could generate an alert to an elderly user with cognitive decline about the forthcoming appearance of the cleaner.

Person tracking in the context of Smart Homes suffers from a series of technical and social limitations. At the same time these limitations can be interpreted as open research problems:

- Integration of the infrastructure: audio-visual person tracking can potentially require tens of sensors for each of the monitored rooms. Given the absence of standardisation for these technologies it is still difficult to predict the exact set of sensors required, their minimum setup and their interconnection to the processing unit. Even though modern buildings include connectivity for Internet Protocol devices, it is not clear whether these networks can support the bandwidth required or how someone can hide the sensors and the corresponding technology clutter. The problems of course become even greater when considering old buildings.
- Computational power: As computers become more powerful the computer power required for person tracking in a large home could become a trivial problem. Today though full-scale employment of a tracking system would require a set of computers that create their own space and cost considerations. Two trends attempt to alleviate the computational cost of tracking systems. Firstly, the massive parallel capabilities of modern graphics cards are exploited

[Kirk and Hwu (2010)]. Secondly, cameras are being embedded in DSP boards on which major parts of the detection and tracking algorithms are run [BeagleBoard (2009)], whilst a single computer would simply fuse the decisions. This approach also alleviates the burden of transferring the images themselves. This is becoming increasingly important as high resolution multi-camera systems are deployed.

- Occlusions and lighting conditions: A home environment differs vastly from the idealised conditions of a research lab. Looking first into the purely algorithmic challenges of video tracking we can identify the fact that people will often sit close to each other creating occlusions and merging tracks. Additionally, adaptive versions of the algorithms have to take into account the varying lighting conditions due to lights and/or time of day. Both video and audio components can obviously benefit from the multi modal version of the tracker. In dark or visually cluttered scenarios the audio module could be trusted more. In cases where noise or cross talk appear to create acoustic conditions that do not provide accurate estimates, the video module would take over. Infrared cameras could prove a viable solution for video tracking during very dark conditions.

- Noise and acoustics: On the audio side algorithmic challenges are created when considering the facts of the noise created by appliances and entertainment devices that become active in a random manner. Additionally, the acoustics of each room will affect performance significantly. Solutions to these problems could be achieved by creating acoustic maps of each enclosure. In this way the system could integrate knowledge about the location of noise sources (e.g. air-conditioning) or a piece of furniture and prefilter the recorded audio signals prior to their use for ASL.

- Acceptance: One of the greatest challenges faced with all flavours of the UC paradigm is its social acceptance. People do not want their houses to turn into monitoring devices that can potentially jeopardise their privacy or the peace provided in a domestic environment. As much as engineers like to detach their research from social implications, person tracking is a technology that ought to take into account the environment in which it is employed.

6.1.2   *The Smart Office*

A Smart Office is another typical example of the UC paradigm. Given the corporate flavour of the application a Smart Office allows for easier integration of pervasive devices, sensors, computing hardware and software in order to realise a context-aware meeting room or a working environment that senses ongoing activities and helps people become more productive.

When compared to Smart Homes, research for Smart Offices has gained more attention since most research laboratories can replicate easier the particularities of an office space. Most of these efforts concentrate on everyday scenarios such as the facilitation of meetings and lectures. In that respect, a Smart Office should offer solutions to the perceptual problems needed to derive a better recognition and understanding of the human context at work, without which all attempts at autonomy would be futile.

A relatively large project in the area was Computers in the Human Interaction Loop (CHIL) [CHIL (2007); Waibel and Stiefelhagen (2009)]. The project consortium worked with the vision of a general class of systems and services that put human-human interaction at the centre and computing services in a supporting function on the periphery. A large part of the project was devoted to perceptual technologies that can operate in realistic unconstrained environments, without simplifications. Central to these was multiple person tracking during meetings and lectures. The complementarity of different modalities (vision, speech, etc.) was also examined in order to better describe human activities. An interesting categorisation of problems when designing Smart Offices was the "Who? What? Where? To Whom? How?" motto that guided most of the perceptual processing research of the project. In this, person tracking answers the "Where?" question and provides the initial metadata for all the rest. An example of a 3D face tracker providing location metadata in a office meeting scenario is shown in Fig. 6.2. Such metadata are utilised by the CHIL system to provide an abstraction of the meeting/lecture room status and participants. Typical screenshots of the system in action are shown in Fig. 6.3.

A project with similar scope is the Augmented Multiparty Interaction (AMI) project [AMI (2009)]. Again, researchers work with the aim of developing technologies that help people have more productive meetings. The initial phase of the project focused on the concept of meeting browsers; tools that allow users to find the information that they want quickly from a recorded meeting. The consortium then steered its attention to content linking, or finding archived documents and segments of meeting recordings

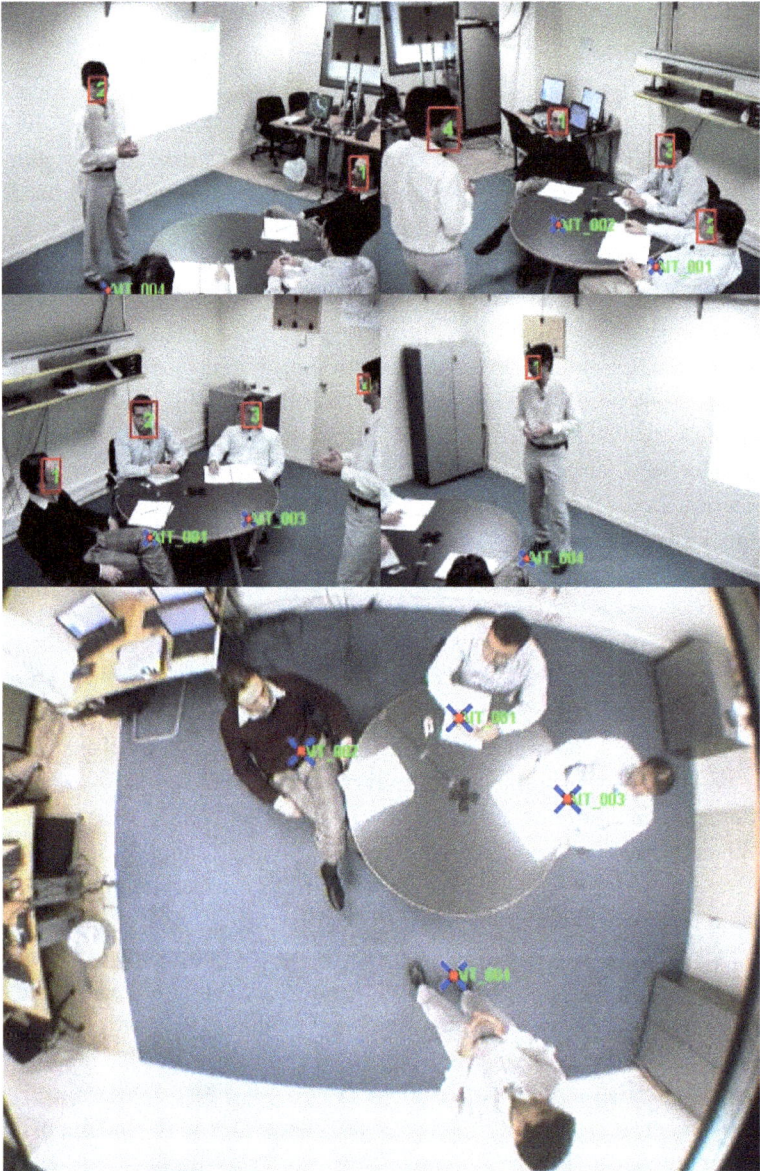

Fig. 6.2   3D tracker providing the locations of all participants in a meeting. The system fuses face detections from the corner cameras (top four frames) and motion evidence from the panoramic camera (bottom frame) to track the projection of the head on the floor [Karame *et al.* (2007); Katsarakis *et al.* (2008)].

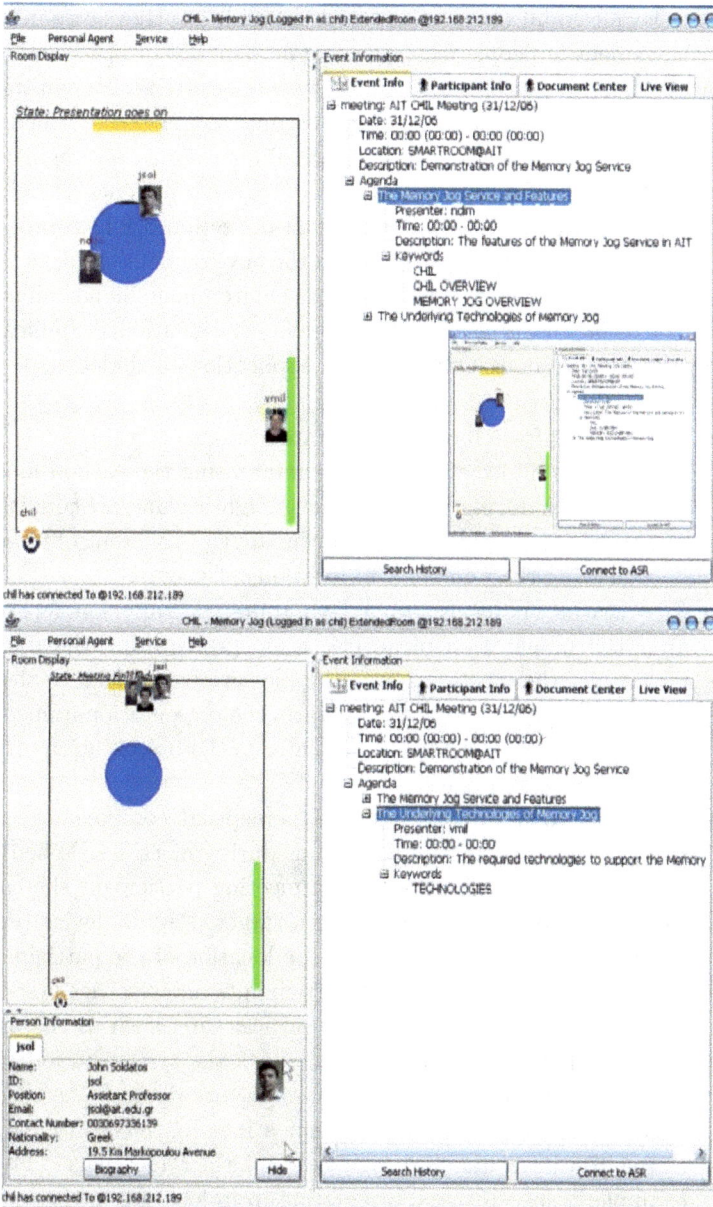

Fig. 6.3 Abstraction of the meeting room: two instances with the meeting in progress (top) and the meeting just finishing and the user querying the system about a participant (bottom).

that are relevant to an ongoing meeting. In addition they attempt to increase engagement of people who are using a telephone or other technology to connect to a meeting that otherwise they cannot attend. Person tracking is mainly used as a pre-processing step in order to offer beamformed audio signals to speech recognisers as well as a facilitator for face recognition.

The applications of person tracking in office environments resemble abstractly those of the Smart Home discussed in section 6.1.1. Applications are service-driven and depend strongly on the environment employed. Nevertheless, person tracking serves all of these environments in a similar manner. Thus except for variations in the previously mentioned applications, an indicative list of Smart Office specific applications includes:

- Given the shifts in modern corporate responsibility, person tracking can serve as an excellent tool for minimising the carbon footprint of companies during working times. Lighting and air-conditioning can be tuned according to the occupancy of different office spaces by observing the number and locations of people.
- Presentations in meetings can be automated. For example, straight after the preparation of an electronic presentation the presenter moves into the meeting room, the files are transferred to the local computer and the Smart Office starts the presentation immediately after the tracker detects the presenter in the presentation area and the audience is in place.
- With the collaboration of other perceptual components, person tracking can minimise interruptions during meetings. When a secretary receives a phone call for a meeting participant she can notify him automatically by asking a targeted-audio device [Pompei (1999)] to send a notification to the location of the participant.
- Knowing the locations of all participants allows other systems to create abstractions of a meeting or a lecture, such as the one depicted in Fig. 6.3. People can then ask the system to identify an unknown meeting participant through any device. At the same time the Smart Office can segment a meeting recording by annotating the interchanging speakers and their locations. Thus for example, a speech signal originating from location in the audience can be interpreted as a question while speech signals from the presenter are annotated as answers. There exists special interest for a specific version of this annotation process called *diarisation*. This involves the identification of the time and identity of speakers dur-

ing a meeting/conversation [Tranter and Reynolds (2006)]. Identifying the locations of the participants can help towards the solution of the diarisation problem [Pardo *et al.* (2007)].

All of the challenges imposed by domestic environments (see section 6.1.1) apply in the corporate case. In fact some of them become even greater. Typically recordings with cameras and microphones in office spaces will suffer from more occlusions and noise while large glass window surfaces will accentuate reverberation problems. Additionally, a person tracker would probably have to comply with some legacy computational or tracking infrastructure. Nevertheless, corporate environments generally exhibit greater acceptance to new technologies and thus Smart Offices could serve as the driving environment for perceptual technologies such as Person Tracking.

6.1.3 *Video-conferencing*

Video-conferencing (VC) is a well-established technology with applications to both Smart Homes and Offices. Nevertheless, due to its maturity we treat this specific application separately in this paragraph. Commonly in current VC systems, the control of the video camera is performed in one of two ways:

- Cameras provide a single or multiple fixed views of the speaker(s) in the room. Placed at different locations in the conference room the cameras are able to cover all the people in it. All video streams are transmitted to the other party.
- Systems consisting of cameras operated by humans. A director decides the camera to be used and the video stream to be transmitted. The resulting video feed is known as the mixed output.

These systems turn out to be expensive in terms of extra hardware or manpower required to operate them effectively and reliably. It would be desirable to have one or two video cameras that can be automatically steered towards the active speaker. Since participants in the video conference might be moving around in the room a person tracking system that reliably locates and tracks the active speaker becomes a necessity. Given that both audio and video feeds are available the speaker localisation and tracking can be performed using both visual or acoustic data.

VC comes in different flavours depending on the environment in which it is used, the number of interconnected parties and the technology available. PC-based VC is the most common version of video-conferencing. It has as minimum prerequisites a single low resolution web-camera, a single microphone and an internet connection through which the conferencing takes place. It became quickly apparent that this minimum setup had its limitations in several scenarios. Ambient noise, cross talk, reverberation and lighting conditions degrade the quality of transmitted signals. Additionally, numerous people prefer to listen to music as they speak but they wish to have this removed prior to the transmission of their voice.

The de-noising and beamforming algorithms required for these tasks quickly equipped the web-cameras with high-resolution video sensors and multiple microphones. Web-cameras can now steer their attention towards the active speaker and improve to a large extent the transmitted audio signal. Given that people move in front of the camera there is the need for employing an ASL system. In most cases tracking the direction of arrival of the speaker is sufficient since the speaker does not move away from camera when using his PC. Some of the VC web-cameras also come equipped with face detection (since speakers face the camera at all times during a VC) in order to track the speaker. This improves the awareness of the system regarding the location of the speaker and beamforming becomes more effective. The system becomes even more accurate with subsequent lip-detection or the use of stereo cameras so that the system has a good estimate of the depth. Multi-modal versions are also available.

Large scale VC with dedicated rooms is common in corporate environments where companies want to bring together employees from different geographical locations into a single virtual meeting space. Installations for such VC rooms vary but for each site there are multiple screens and cameras focusing and displaying participants from all other sites.

Person tracking in such spaces can lower equipment-related costs and minimise the feel of obtrusiveness. A modern large scale VC has no need for close-talking microphones. Multiple microphone arrays on each wall ensure recording of all participants. Redundancy only helps in this case! A multi-modal person tracker can steer cameras towards the active speaker and minimise the need for multiple camera streams from each participating site. Given that these rooms are dedicated to VC purposes they can be designed for minimum technology-clutter while retaining optimal acoustics and lighting so that problems found in non-controlled applications are diminished.

With the explosion of the mobile market VC has became significant even between mobile phone users. Mobile phones now include both cameras and small microphone arrays (two microphones) in an effort to track the location of the speaker and improve the fidelity of the communication. In a slightly different setting, in the challenging environment of vehicles, multi modal person tracking has also became significant. Special car installations with microphone arrays and cameras enable the vehicle to pre-filter speech which is then transmitted during a hands-free conversation or fed into an ASR system for controlling vehicle systems.

6.2 Entertainment

One of the fundamental properties of the physical world that humans exploit is spatial dimensionality. Humans explain dimensionality verbally in terms of three dimensions i.e. the height, width and depth from which sound is arriving. In general humans are able to recognise the dimensionality and directionality of sound very well especially when they invoke for this task both their ears and eyes.

Audio systems have been developed using multiple loudspeakers or headphones in order to reproduce 3D sound, in which the brain perceives the sounds as coming from a particular direction around the head. Special interest has been given to this research direction after the expanding popularity of home entertainment systems that attempt to recreate complex sound-fields using multiple loudspeakers around the listener or more recently using linear loudspeaker arrays. It is worth noting that the reproduction quality is directly connected to the size and shape of the room, as well as to furniture and wall materials.

In order for 3D audio to provide useful relative information and a corresponding realistic experience, the position and orientation of the head of the user with respect to a common reference frame must be available. Early systems assumed that there is a single listener staying still in a single position and require accurate positioning of the loudspeakers. Other systems have examined the use of inertial sensors for head tracking [Ababsa and Mallem (2004)]. This process involves an inertial head-orientation tracker based on gyroscopes, inclinometers and a digital compass. These are integrated into the headphones of the listener or in the case of speakers onto a device that the listener wears on his/her head.

Person tracking could potentially replace the obtrusive nature of inertial tracking. Using the cameras installed in a Smart Home, entertainment systems can gain access to the coordinates of all the people in the room and control directionality of sound. This also allows the system to cope effectively with listeners moving since the coordinates would be available on-line.

Another area that receives lots of attention in the computer games industry is augmented reality. These systems fuse real-world objects into a virtual world, where abstractions of the real-world objects exist. In games, the complete human body or just the hands are modelled into a virtual world to control game characters. To sense the human body and the movement of its parts person trackers are used. The articulated body is transformed into a set of parameters for human body models, which are subsequently rendered in the virtual world.

An example of an augmented reality fun application allows for hands or any other object on or over a table top to interact with virtual cubes [Anagnostopoulos and Pnevmatikakis (2008)]. In order to do so, a physics engine models the objects and the rules of their interaction (gravity, friction, percussion). The actual objects are inserted into the virtual world of the physics engine by means of a 3D visual tracker. The tracker utilises a stereo camera pair. The measurement process segments all non-grey objects and if these match a palm, models the palm in a 13-dimensional state-space: two dimensions for the palm centre, one for the primary axis angle and two for the polar coordinates of every fingertip . The average height of the segmented object is estimated from the disparity between the right and the left images of the stereo camera. The application in action is shown in Fig. 6.4.

6.3   Personalised media broadcasting

Nowadays there is a trend towards personalised delivery of multimedia streams, with applications in video-on-demand and personalised coverage of sports events. A platform that empowers viewers to direct their own coverage of large athletics events should enable them to set up their own personal virtual director to manage camera coverage of an event and orchestrate event viewing to their own preferences. For example, broadcasting of athletics tends to be biased towards track events [Poslad *et al.* (2009)]. Personalised content delivery can change that for viewers who wish to focus

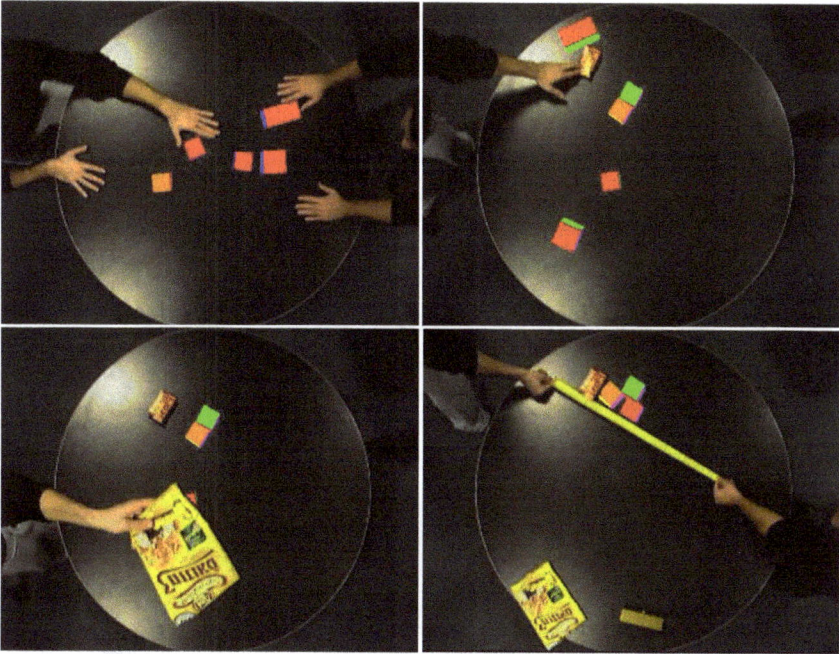

Fig. 6.4 The augmented reality application. (Top left) Hands interacting with the virtual objects. (Top right) Real objects interacting with virtual objects. (Bottom left) Real objects occluding virtual objects. (Bottom right) Real objects interacting with a mix of real and virtual objects: the real and virtual objects are dragged off the table.

on different content. Such a platform employs video processing to extract automatic video annotations that are used for reasoning about incidents of interest. Subsequently, viewers' preferences are matched to the detected incidents in order to issue recommendations on which camera to view. Finally dynamic adaptation to the end-user environment, including terminal and network capabilities is also important.

There are two types of camera proposals for personalised content delivery. In the case of between-sports camera selection, the view is changed between two sports, upon occurance of an incident in a sport higher up the user preferences than the one currently being delivered. In within-sport camera selection, the camera is changed to offer a better view of the evolution of the sport, or the incident of interest [e Director 2012 (2010)].

Automatic video annotation includes tracking the way the cameras of the broadcaster move and tracking the athletes as they compete [Pnev-matikakis *et al.* (2010)]. All the algorithms in Chapter 4 can be utilised.

Sound from sports venues is also used, since crowd excitement is an important cue for an incident of interest, but audio tracking is not attempted since there is no individual source to track.

An example of incident detection in long jump utilising face tracks from a frontal camera and camera shift from a profile camera is shown in Fig. 6.5. The detected incidents are compared to the manual annotations provided by a sport information system, typically employed in large scale sports events.

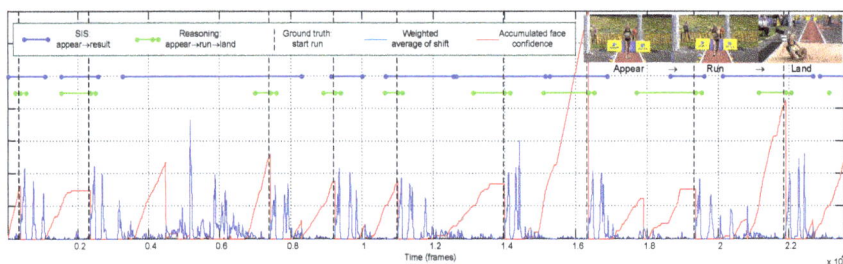

Fig. 6.5 Automatic annotation of the phases in a long jump event: the vision processing metrics used are the accumulated face track likelihood (red) and the weighted average of camera shift (blue). The resulting incident reasoning about athlete presence, start or run and land is depicted by green horizontal bars. Typical screenshots of the three phases are also overlaid in the figure. The manual annotations from appearance of athlete to distance of jump measured are depicted in red. The ground truth as the actual moment of beginning of a run is marked by the vertical dashed black lines.

6.4   Security and surveillance

Perhaps the oldest application of person tracking is for surveillance in secure spaces. Surveillance systems aim to monitor the tracks of people in secure spaces. They range from simple dead zone monitoring systems, to complicated behaviour analysis systems. The dead zone monitoring systems just sense the presence of humans where there should be none. They are more complicated than simple motion sensors, especially for outdoor environments where motion can be many things other than unauthorised access. The behaviour analysis systems attempt to recognise the intentions of people in the monitored space. Group meeting and loitering are typical behaviours of interest and the monitored spaces are typically spaces where human activity is expected and common, such as transport hubs. An example of an outdoor surveillance system employing motion-based pedestrian

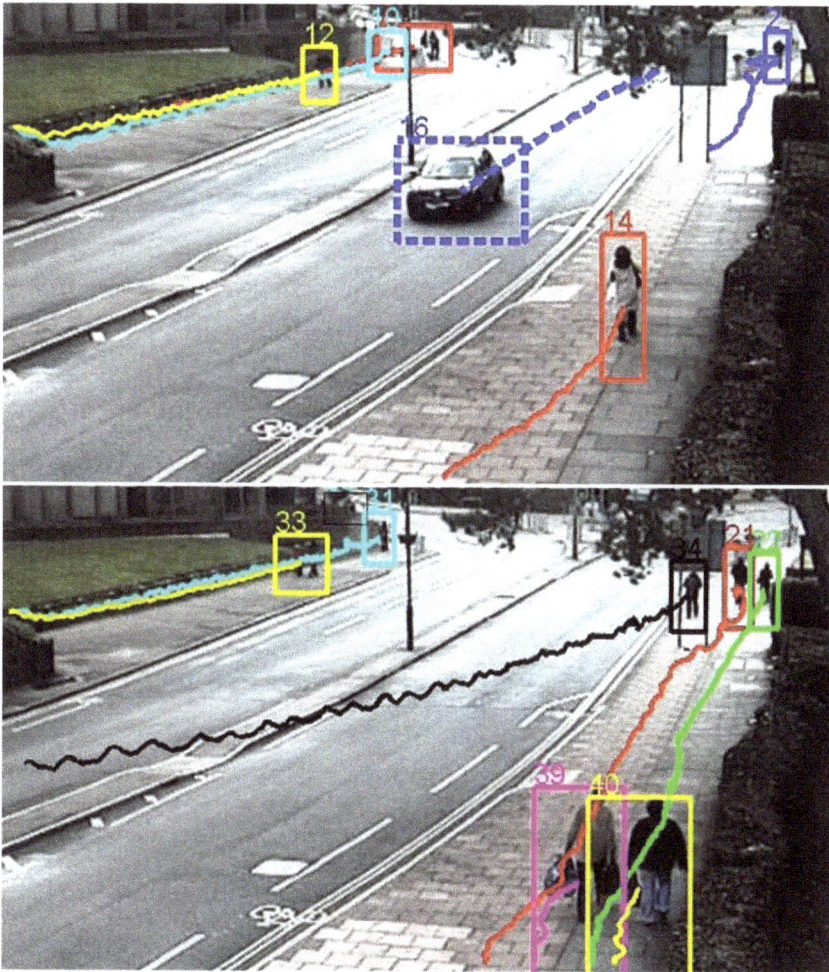

Fig. 6.6 Outdoor surveillance systems. Targets are recognised as pedestrians (solid bounding boxes) or vehicles (dashed bounding boxes).

and vehicle trackers [Stergiou *et al.* (2007)] is shown in Fig. 6.6. It is built around the motion-based tracker detailed in section 4.11.1.2.

Due to the lack of microphones (in most countries it is not allowed to monitor spaces with microphones) audio or multi modal tracking has almost no role in surveillance applications. Additionally, the diverse conditions found outdoors, the person clutter and the quite far-field nature of the recordings complicate tracking. As well as person tracking, surveillance systems also need track analysis and camera hand-over modules. The former is for behaviour analysis, while the latter allows individuals to be tracked across cameras in camera networks even when there is no common field of view between the individual cameras.

6.5   Robotics

Research in the area of androids or simpler autonomous robots, has been popular and active for several decades in the Artificial Intelligence community. These efforts uncovered a variety of new problems in robotics that extended far beyond the replication of human movements. Modern robotics attempts to replicate all human senses in order to give robots a complete understanding of their surroundings. An understanding that is then processed and fused to lead to decision making and possibly to a level of intelligence.

An analytical review of modern robotics is out of the scope of this book. Nevertheless, it is worth mentioning a few exemplary systems that connect to different extents to person tracking. In the past decades robots have moved from being remotely controlled devices to autonomous systems equipped with improved robot-human interfaces.

The Japan Science and Technology Agency has developed a new humanoid robot "Child-robot with Biomimetic Body (CB2)" [Minato *et al.* (2007)] that has a whole-body soft skin (a silicon surface with 197 high sensitive tactile sensors underneath) and flexible joints (51 pneumatic actuators). Vision is realised by cameras mounted inside the eyeballs. Audition is realised by microphones mounted both on the head, and located externally, around the head of CB2.

With a similar set of sensors scientists have created the Philip K. Dick android [PKDandroid (2006)]. Dick was a science fiction writer who explored what it means to be human, and wrote extensively about humans, androids and identity. The robot operated without human intervention. It tracked people coming in and out of the room with face recognition software, and would greet faces that it knew. It listened to verbal input and used complex algorithms to generate a response and then responded verbally using speech synthesis. The project was abruptly terminated when the android went missing on a flight in early 2006.

The Kismet of MIT AI Lab [Breazeal (2000)] can recognise speech by a speech-recognition system and express various kinds of sensation. All audio signals used are pre-filtered using the location of the speaker. In the same manner the Hadaly robot of Waseda University [Hashimoto *et al.* (2002)] can localise the speaker as well as recognise speech by a speech-recognition system.

The ASIMO service and companion robot by HONDA [Honda (2008)] can recognise moving objects, faces and gestures, can chart a route and can distinguish sounds and their direction of arrival. This is achieved by fusing measurements from a variety of sensors, including five microphones, a stereo camera pair for the head, ultrasonic sensors for the surrounding three metres, and finally a laser sensor and an infrared sensor for the immediate floor.

There are many more efforts from scientists around the world to design robots that can integrate into human everyday life. It has became apparent that such robots should be able to sense the environment in a way similar to humans. Thus, a large part of the research that concerns the replication of human audio-visual tracking has been detached from robotics and treated separately. For auditory and visual tracking, a characteristic example can be found in [Nakadai *et al.* (2000)] where authors presented an active audition system for humanoids in order to improve sound source tracking by integrating audition, vision, and motor controls. An active audition system is implemented in an upper-torso humanoid to demonstrate that the humanoid actively moves its head to improve localisation by aligning microphones orthogonal to the sound source and by capturing the possible sound sources by vision.

The algorithms and systems presented in earlier chapters of this book could have a direct application in a robot or humanoid. Employing a minimum of two microphones can allow for direction of arrival estimation and subsequent beamforming and/or denoising. The filtered data can then be used to feed a context extraction system (e.g. an ASR) in order to enable a conversation. Alternatively, the robot can simply follow a specific path by identifying the location from which sound is arriving.

A video-based person tracking system has an equally important role. Using a stereo camera, the exact location of people around the robot can be identified and tracked. Tracking people can prove useful in a series of applications such as face-recognition.

Combination of audio and video systems can allow the robot to operate in dark or noisy environments but most importantly it allows it to model more easily its surrounding environment and to initiate a learning process. Tracking and identifying people is a natural process for humans. Thus, it is expected that robots supporting such abilities will integrate into human societies quicker, more naturally and most importantly with higher acceptability.

6.6   Marketing and retail

The effectiveness of advertising is traditionally measured in terms of cost per thousand. This is computed by dividing the cost of advertising with the audience measure. In order to improve the market viability of Out-Of-Home (OOH) advertising, a system is required for also improving the method of measuring audience metrics and, ultimately, its effectiveness. This requires a system that more accurately measures the impact of the ad through prescribed measures (presence, notice, dwell time), a system that detects and measures actual advertisment viewing or the time spent in a specific area (e.g. at an expo kiosk).

Person tracking can provide such an effective breakthrough in audience measurement. Video-based localisation and tracking can automatically provide instantaneous audience data for immediate use and for consolidation by any data-mining or statistical analysis service. For digital-signage operators to content providers, person tracking can potentially offer an audience measurement solution which is easy to deploy by using cameras and tailored versions of the trackers. Using a standard video sensor in the vicinity of the measured media and pointing it at the intended audience the system can then analyse the video stream to detect the presence of human faces, track those faces while they remain in the field of vision and finally classify those faces in anonymous demographic classes. Alternatively a ceiling mounted camera can measure the time statistics of a specific region i.e. how much time people spend at a specific area in the vicinity of the camera. An example of a system measuring the coverage that moving people have generated on the floor is shown in Fig. 6.7. It is built around the motion-based body tracker detailed in section 4.11.1.2.

The in-shop behaviour of clients is very important for retailers. Security is just a small aspect of such systems. What is of more interest is the pattern of the movement of clients in stores, especially in large ones. Reactions of the clients to the various promotional activities is also important. Such information is collected by track analysis systems, which are based on person trackers. An example of a system analysing tracks in a bookstore is shown in Fig. 6.8. Track features of interest in such systems are length,

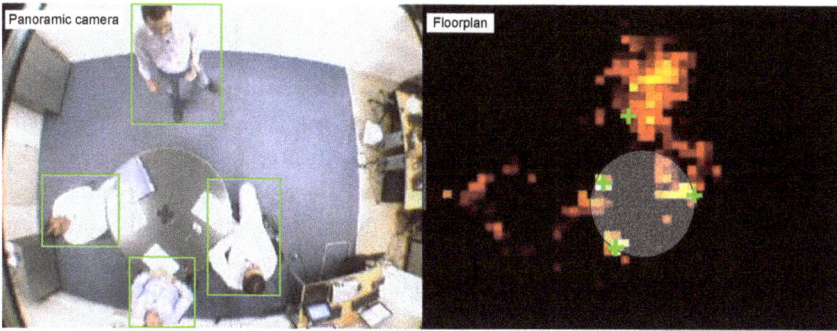

Fig. 6.7 Targets tracked from a panoramic camera (left) and resulting coverage on the floor (right). Brighter colours indicate highest occupancy at the particular 10-by-10 cm square.

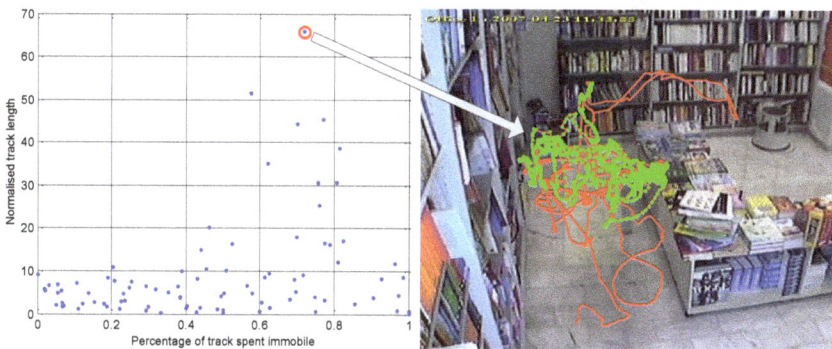

Fig. 6.8 Analysis of tracks in a bookstore: tracks are represented by their normalised length and their duration fraction during which the target is almost immobile (left). The longest and most immobile track is shown in red, with the part where the person is almost immobile being overlaid in green (right). That client, upon reaching a particular bookshelf in the store, spent quite a lot of time almost immobile browsing books, hence is an interested client.

the portion of the store covered, speed in the various sections of the store and the points of focus.

Using audience measurement and client behaviour solutions, manufacturers and retailers can measure the number of viewers and the level of their interest when advertising on digital signs and television monitors, and the interest their displays and in-shop promotion activities have generated.

6.7   Conclusions

Person tracking systems have entered our lives. All of us know that they are out there to deter unwanted behaviour, but this is only a small part of their usage. They are now begining to facilitate the way we work, they ease the life of sensitive population groups, they prolong independence of the elderly, they allow us to be more environmentally-friendly and they entertain us.

In the future, such systems will allow technology to fade into the background and will anticipate human commands and intentions, as they are expressed in the ways natural to humans. Human-centric services can thus be launched, taking into account our location, activity and even mood.

Bibliography

AAL (2010). The ambient assisted living joint programme, http://www.aal-europe.eu/about-aal.

Ababsa, F. and Mallem, M. (2004). Inertial and vision head tracker sensor fusion using a particle filter for augmented reality systems, in *Proceedings of the 2004 International Symposium on Circuits and Systems (ISCAS '04)*, Vol. 3 (Vancouver, Canada), pp. 861–864.

Allen, J. and Berkley, D. (1979). Image method for efficiently simulating small-room acoustics, *Journal of Acoustical Society of America* **65**, 4, pp. 943–950.

AMI (2009). Augmented Multi-party Interaction, http://www.amiproject.org/.

Anagnostopoulos, A. and Pnevmatikakis, A. (2008). A realtime mixed reality system for seamless interaction between real and virtual objects, in *DIMEA '08: Proceedings of the 3rd international conference on Digital Interactive Media in Entertainment and Arts* (Athens, Greece), pp. 199–204.

Andersen, M., Andersen, R., Katsarakis, N., Pnevmatikakis, A. and Tan, Z.-H. (2010). Three-dimensional adaptive sensing of people in a multi-camera setup, in *Person tracking for assistive working and living environments, EUSIPCO 2010* (Aalborg, Denmark), pp. 964–968.

Arulampalam, S., Maskell, S., Gordon, N. and Clapp, T. (2002). A tutorial on particle filters for on-line non-linear/non-gaussian bayesian tracking, *IEEE Transactions on Signal Processing* **50**, 2, pp. 174–188.

Azodolmolky, S., Dimakis, N. and Mylonakis, V. (2005). Middleware for indoor ambient intelligence: The polyomaton system, in *Proceedings of the 2nd international workshop on Next Generation Networking Middleware (in conjunction with Networking 2005)* (Waterloo, Canada), pp. 943–950.

Babu, R. V., Pérez, P. and Bouthemy, P. (2007). Robust tracking with motion estimation and local kernel-based color modeling, *Image and Vision Computing* **25**, 8, pp. 1205–1216.

Badrinarayanan, V., Pérez, P., Clerc, F. L. and Oisel, L. (2007). Probabilistic color and adaptive multi-feature tracking with dynamically switched priority between cues, in *IEEE 11th International Conference on Computer Vision* (Rio de Janeiro, Brasil), pp. 1–8.

BeagleBoard (2009). Beagleboard: System reference manual, http://beagleboard.org/static/BBSRM_latest.pdf.

Beal, M., Attias, H. and Jojic, N. (2002). Audio-video sensor fusion with probabilistic graphical models, in A. Heyden, G. Sparr, M. Nielsen and P. Johansen (eds.), *Computer Vision ?ECCV 2002, Lecture Notes in Computer Science*, Vol. 2350 (Springer, Berlin/Heidelberg, Germany), pp. 736–750.

Benesty, J., Huang, Y. and Chen, J. (2007). Time delay estimation via minimum entropy, *IEEE Signal Processing Letters* **14**, 3, pp. 157–160.

Bernardin, K., Gehrig, T. and Stiefelhagen, R. (2008). Multi-level particle filter fusion of features and cues for audio-visual person tracking, in R. Stiefelhagen, R. Bowers and J. Fiscus (eds.), *Multimodal Technologies for Perception of Humans, Lecture Notes in Computer Science*, Vol. 4625 (Springer, Berlin/Heidelberg, Germany), pp. 70–81.

Bernardin, K., Stiefelhagen, R., Pnevmatikakis, A., Lanz, O., Brutti, A., Casas, J. R. and Potamianos, G. (2009). Person tracking, in A. Waibel and R. Stiefelhagen (eds.), *Computers in the Human Interaction Loop* (Springer-Verlag, London, UK), pp. 11–22.

Blackman, S. (1986). *Multiple-target tracking with radar applications* (Dedham, MA, Artech House, Inc.).

Blackman, S. and Popoli, R. (1999). *Design and Analysis of Modern Tracking Systems* (Artech House radar library).

Blake, A. and Isard, M. (1998). *Active Contours: The Application of Techniques from Graphics, Vision, Control Theory and Statistics to Visual Tracking of Shapes in Motion* (Springer-Verlag New York, Inc., Secaucus, NJ, USA).

Bolic, M., Djuric, P. and Hong, S. (2002). Source localization and beamforming, *IEEE Signal Processing Magazine* **19**, 2, pp. 30–40.

Bolic, M., Djuric, P. and Hong, S. (2003). New resampling algorithms for particle filters, in *Proceedings of the IEEE International Conference on Acoustics, Speech, and Signal Processing*, Vol. 2 (Hong Kong, China), pp. 589–592.

Bouguet, J.-Y. (2008). Camera calibration toolbox for matlab, www.vision.caltech.edu/bouguetj/calib_doc/htmls/parameters.html.

Bradski, G. (1998). Computer vision face tracking for use in a perceptual user interface, in *Intel Technology Journal*, Vol. 2.

Bradski, G., Kaehler, A. and Pisarevsky, V. (2005). Learning-based computer vision with intel's open source computer vision library, in *Intel Technology Journal*, Vol. 9, pp. 119–130.

Brandstein, M. and Ward, D. (eds.) (2001). *Microphone Arrays Signal Processing Techniques and Applications* (Springer, London, UK).

Brandstein, M. S., Adcock, J. E. and Silverman, H. F. (1997). A closed-form location estimator for use with room environment microphone arrays, *IEEE Transactions on Acoustics Speech and Signal Processing* **5**, pp. 45–50.

Breazeal, C. (2000). Kismet, the robot, http://groups.csail.mit.edu/lbr/sociable/baby-bits.html.

Brumitt, B., Meyers, B., Krumm, J., Kern, A. and Shafer, S. (2000). EasyLiving: Technologies for intelligent environments, in P. Thomas and H.-W. Gellersen (eds.), *Handheld and Ubiquitous Computing, Lecture Notes in*

Computer Science, Vol. 1927 (Springer, Berlin/Heidelberg, Germany), pp. 97–119.

Buehren, M. (2009). Functions for the rectangular assignment problem, http://www.mathworks.com/matlabcentral/fileexchange/ 6543-functions-for-the-rectangular-assignment-problem.

Build-IT (2010). Independent living, http://www.build-it.gr/Portals/0/ACIA.Gallery/brochure06EN_Independent_Living.pdf.

Chang, P. and Krumm, J. (1999). Object recognition with color cooccurrence histograms, in *IEEE Conference on Computer Vision and Pattern Recognition (CVPR '99)* (Fort Collins, CO, USA).

CHIL (2007). CHIL (Computers in the Human Interaction Loop) EU FP6 integrated project, http://chil.server.de/.

Comaniciu, D., Ramesh, V. and Meer, P. (2000). Real-time tracking of non-rigid objects using mean shift, in *IEEE Computer Society Conference on Computer Vision and Pattern Recognition (CVPR'00)*, Vol. 2 (Hilton Head, South Carolina, USA), p. 2142.

Cour, B. L. (2007). Ensemble-based Bayesian detection and tracking, *The Journal of the Acoustical Society of America* **121**, 5, pp. 3191–3191.

Cover, T. and Thomas, J. (1991). *Elements Of Information Theory* (Wiley, London, UK).

Crowley, J. L. (2003). Context driven observation of human activity, in *Proceedings of the European Symposium on Ambient Intelligence* (Veldhoven, The Netherlands), pp. 101–118.

Dalley, G. (2009). Videoio toolbox for matlab, http://videoio.sourceforge.net.

Deller, J., Proakis, J. and Hansen, J. (1993). *Discrete-Time Processing of Speech Signals* (Macmillan Publishing Company, London, UK).

Dey, A. K. (2001a). Person location and person tracking - technologies, risks and policy implications, *Information Technology and People* **14**, pp. 206–231.

Dey, A. K. (2001b). Understanding and using context, *Personal and Ubiquitous Computing* **5**, pp. 4–7.

Douc, R., Cappé, O. and Moulines, E. (2005). Comparison of resampling schemes for particle filtering, in *4th International Symposium on Image and Signal Processing and Analysis (ISPA)* (Zagreb, Croatia), pp. 64–69.

Doucet, A., Godsill, S. and Andrieu, C. (2000). On sequential monte carlo sampling methods for bayesian filtering, *Statistics and Computing* **10**, 3, pp. 197–208.

e Director 2012, M. (2010). My-e-Director 2012 (real-time context-aware and personalized media streaming environments for large scale broadcasting applications) EU FP7 specific targeted research project, http://www.myedirector2012.eu.

Forsyth, D. and Ponce, J. (2002). *Computer Vision: A Modern Approach* (Prentice Hall, Upper Saddle River, NJ, USA).

Fujifilm (2006). Fujifilm face detection, http://www.letsgodigital.org/en/9024/fujifilm_face_detection/.

Funt, B. V. and Finlayson, G. D. (1995). Color constant color indexing, *IEEE Transactions on Pattern Analysis and Machine Intelligence* **17**, 5, pp. 522–529.

Gatica-Perez, D., Lathoud, G., McCowan, I. A., Odobez, J.-M. and Moore, D. (2003). Audio-visual speaker tracking with importance particle filters, in *IEEE International Conference on Image Processing (ICIP)* (Barcelona, Spain), pp. 25–28.

Gaubitch, N. (2003). An implementation of the image model for small room acoustics, http://www.commsp.ee.ic.ac.uk/ ndg/.

Gibson, W. (1992). *Neuromancer* (Heyne, Verlang, München, Germany).

Gilat, A. (2004). *MATLAB: An Introduction with Applications* (John Wiley, New York, USA).

Gonzalez, R. C. and Woods, R. E. (2007). *Digital Image Processing*, 3rd edn. (Prentice Hall, Upper Saddle River, NJ, USA).

Gordon, N., Salmond, D. and Smith, A. (1993). Novel approach to nonlinear/non-Gaussian Bayesian state estimation, *IEE Proceedings-F (Radar and Signal Processing)* **140**, 2, pp. 107–113.

Hashimoto, S., Narita, S. and Kasahara, H. (2002). Humanoid robots in waseda university?adaly-2 and wabian, *Autonomous Robots* **12**, pp. 25–38.

HERMES (2010). HERMES (cognitive care and guidance for active ageing) EU FP7 specific targeted research project, http://www.fp7-hermes.eu.

Honda (2008). Inside ASIMO, http://asimo.honda.com/InsideASIMO.aspx.

Hsu, R.-L., Abdel-Mottaleb, M. and Jain, A. K. (2002). Face detection in color images, *IEEE Transactions on Pattern Analysis and Machine Intelligence* **24**, 5, pp. 696–706.

Hua, G. and Wu, Y. (2006). Measurement integration under inconsistency for robust tracking, in *2006 IEEE Computer Society Conference on Computer Vision and Pattern Recognition* (New York, USA), pp. 650–657.

Isard, M. and Blake, A. (1998). CONDENSATION - conditional density propagation for visual tracking, *International Journal of Computer Vision* **29**, 1, pp. 5–28.

Jaffré, G. and Crouzil, A. (2003). Non-rigid object localization from color model using mean shift, in *IEEE International Conference on Image Processing (ICIP 2003)* (Barcelona, Spain), pp. 317–320.

Jang, D.-S., Jang, S.-W. and Choi, H.-I. (2002). 2D human body tracking with structural Kalman filter, *Pattern Recognition* **35**, 10, pp. 2041–2049.

Jones, M. J. and Rehg, J. M. (2002). Statistical color models with application to skin detection, *International Journal of Computer Vision* **46**, 1, pp. 81–96.

Kalman, R. E. (1960). A new approach to linear filtering and prediction problems, *Transactions of the ASME–Journal of Basic Engineering* **82**, Series D, pp. 35–45.

Karame, G., Stergiou, A., Katsarakis, N., Papageorgiou, P. and Pnevmatikakis, A. (2007). 2D and 3D face localization for complex scenes, in *IEEE International Conference on Advanced Video and Signal based Surveillance (AVSS 2007)* (London, UK), pp. 371–376.

Katsarakis, N. and Pnevmatikakis, A. (2009). Face validation using 3d information from single calibrated camera, in *Proceedings of the 16th international
conference on Digital Signal Processing (DSP'09)* (Santorini, Greece), pp.
972–977.

Katsarakis, N., Talantzis, F., Pnevmatikakis, A. and Polymenakos, L. (2008).
The AIT 3D audio / visual person tracker for clear 2007, in R. Stiefelhagen, R. Bowers and J. Fiscus (eds.), *Multimodal Technologies for Perception of Humans, Lecture Notes in Computer Science*, Vol. 4625 (Springer,
Berlin/Heidelberg, Germany), pp. 35–46.

Kay, S. (1993). *Fundamentals of Statistical Signal Processing: Estimation Theory*
(Prentice-Hall, Upper Saddle River, NJ, USA).

Kientz, J. A., Patel, S. N., Jones, B., Price, M., E., D., E. and Abowd, G. D.
(2008). The Georgia Tech Aware Home, in *26th ACM Conference on Human Factors in Computing (CHI '08)* (Florence, Italy).

Kirk, D. and Hwu, W.-M. (2010). *Programming Massively Parallel Processors: A
Hands-on Approach* (Morgan Kaufmann, Mountain View, USA).

Kitagawa, G. (1996). Monte carlo filter and smoother for non-gaussian nonlinear
state space models, *Journal of Computational and Graphical Statistics* **5**,
1, pp. 1–25.

Knapp, C. and Carter, G. (1976). The generalized correlation method for estimation of time delay, *IEEE Transaction on Acoustics Speech and Signal
Processing* **24**, 4, pp. 320–327.

Koh, M.-S. and Mortz, M. (2000). Improved voice activity detection of noisy
speech, *The Journal of the Acoustical Society of America* **107**, 5, pp. 2907–
2908.

Kwok, C., Fox, D. and Meila, M. (2004). Real-time particle filters, *Proceedings of
the IEEE* **92**, 3, pp. 469–484.

Larson, K. and Topping, R. (2003). PlaceLab: A House_n + TIAX initiative,
http://architecture.mit.edu/house_n/documents/PlaceLab.pdf.

Lehmann, E. and Johansson, A. (2007). Particle filter with integrated voice activity detection for acoustic source tracking, *EURASIP Journal on Advances
in Signal Processing* **2007**, pp. 28–39.

Leichter, I., Lindenbaum, M. and Rivlin, E. (2006). A general framework for combining visual trackers — the "black boxes" approach, *International Journal
of Computer Vision* **67**, 3, pp. 343–363.

Li, S. Z. and Zhang, Z. (2004). Floatboost learning and statistical face detection,
IEEE Transactions on Pattern Analysis and Machine Intelligence **26**, 9,
pp. 1112–1123.

Lucas, B. D. and Kanade, T. (1981). An iterative image registration technique
with an application to stereo vision, in *7th International Joint Conference
on Artificial Intelligence (IJCAI '81)* (Vancouver, BC, Canada), pp. 674–
679.

MacCormick, J. (2002). *Stochastic algorithms for visual tracking* (Springer-
Verlag, London, UK).

MacCormick, J. and Blake, A. (2000). A probabilistic exclusion principle for tracking multiple objects, *International Journal of Computer Vision* **39**, 1, pp. 57–71.

Mathworks (2010). Matlab - the language of technical computing, http://www.mathworks.com/products/matlab/.

Mikolajczyk, K., Choudhury, R. and Schmid, C. (2001). Face detection in a video sequence - a temporal approach, in *IEEE Computer Society Conference on Computer Vision and Pattern Recognition (CVPR 2001)* (Kauai, USA), pp. 96–101.

Minato, T., Yoshikawa, Y., Noda, T., Ikemoto, S., Ishiguro, H. and Asada, M. (2007). CB2: A child robot with biomimetic body for cognitive developmental robotics, in *7th IEEE-RAS International Conference on Humanoid Robots* (Pittsburgh, USA), pp. 557–562.

Moghaddam, B. (2002). Principal manifolds and probabilistic subspaces for visual recognition, *IEEE Transactions on Pattern Analysis and Machine Intelligence* **24**, 6, pp. 780–788.

Molhave, T. (2000). Transformations in 3D, home10.inet.tele.dk/moelhave/ tutors/3d/transformations/transformations.html.

Nakadai, K., Lourens, T., Okuno, H. G. and Kitano, H. (2000). Active audition for humanoid, in *17th National Conference on Artificial Intelligence (AAAI-2000)* (Austin, USA), pp. 832–839.

Nelson, P. and Elliott, S. (1992). *Active Control of Sound* (Academic Press, Maryland Heights, USA).

Nikon (2005). Nikon face-priority AF, http://www.dpreview.com/news/0502/05021604nikonfaceaf.asp.

Nishiura, T., Gruhn, R. and Nakamura, S. (2001). Automatic steering of microphone array and video camera toward multi-lingual tele-conference through speech-to-speech translation, in *IEEE International Conference on Multimedia and Expo (ICME 2001)* (Tokyo, Japan), pp. 447–450.

Orderud, F. (2005). Comparison of Kalman filter estimation approaches for state space models with nonlinear measurements, in *Scandinavian Conference on Simulation and Modeling (SIMS 2005)* (Trondheim, Norway).

Orwell, G. and Crick, B. (1984). *Nineteen eighty-four* (Oxford University Press, Oxford, UK).

Pardo, J., Anguera, X. and Wooters, C. (2007). Speaker diarization for multiple-distant-microphone meetings using several sources of information, *IEEE Transactions on Computers* **56**, pp. 1212–1224.

Peled, Y. and Rafaely, B. (2008). Study of speech intelligibility in noisy enclosures using optimal spherical beamforming, *IEEE 25th Convention of Electrical and Electronics Engineers in Israel* , pp. 285–289.

Pérez, P., Hue, C., Vermaak, J. and Gangnet, M. (2002). Color-based probabilistic tracking, in *7th European Conference on Computer Vision (ECCV 2002)*, Vol. 1 (Copenhagen, Denmark), pp. 661–675.

Pérez, P., Vermaak, J. and Blake, A. (2004). Data fusion for visual tracking with particles, *Proceedings of IEEE* **92**, 3, pp. 495–513.

Peterson, P. M. (1986). Simulating the response of multiple microphones to a single acoustic source in a reverberant room, *The Journal of the Acoustical Society of America* **80**, 5, pp. 1527–1529.

Petsatodis, T., Pnevmatikakis, A., Talantzis, F. and Diaz, U. (2009). Interactive surfaces for enhanced cognitive care, in *16th international conference on Digital Signal Processing (DSP'09)* (Santorini, Greece), pp. 501–504.

PKDandroid (2006). PKD Android: A robotic portrait, http://www.pkdandroid.org/.

Pnevmatikakis, A., Katsarakis, N., Chippendale, P., Andreatta, C., Messelodi, C., S. Modena and Tobia, F. (2010). Tracking for context extraction in athletic events, in *Workshop on Social, Adaptive and Personalized Multimedia Interaction and Access (SAPMIA 2010), ACM Multimedia* (Florence, Italy), pp. 67–72.

Pnevmatikakis, A. and Polymenakos, L. (2006). Robust estimation of background for fixed cameras, in *15th International Conference on Computing (CIC '06)* (IEEE Computer Society, Mexico City, Mexico), pp. 37–42.

Pnevmatikakis, A., Soldatos, J., Talantzis, F. and Polymenakos, L. (2009). Robust multimodal audio-visual processing for advanced context awareness in smart spaces, *Personal and Ubiquitous Computing* **13**, 1, pp. 3–14.

Pnevmatikakis, A. and Talantzis, F. (2010). Person tracking in enhanced cognitive care: A particle filtering approach, in *Person tracking for assistive working and living environments, EUSIPCO 2010* (Aalborg, Denmark), pp. 969–973.

Pompei, F. J. (1999). The use of airborne ultrasonics for generating audible sound beams, *Journal of the Audio Engineering Society* **47**, 9, pp. 726–731.

Poslad, S., Pnevmatikakis, A., Nunes, M., Ostermann, E., Chippendale, P., Brightwell, P. and Patrikakis, C. (2009). Directing your own live and interactive sports channel, in *10th International Workshop on Image Analysis for Multimedia Interactive Services (WIAMIS 2009)* (London, UK), pp. 275–279.

Prasad, R., Muralishankar, R., Vijay, S., Shankar, H., Pawelczak, P. and Niemegeers, I. (2006). Voice activity detection for VoIP – an information theoretic approach, in *IEEE Global Telecom Conference (GLOBECOM)* (San Francisco, USA), pp. 1–6.

Radlović, B. D., Williamson, R. C. and Kennedy, R. A. (2000). Equalization in an acoustic reverberant environment: Robustness results, *IEEE Transactions on Speech and Audio Processing* **8**, pp. 311–319.

Ristic, B., Arulampalam, S. and Gordon, N. (2004). *Beyond the Kalman Filter: Particle Filters for Tracking Applications* (Artech House Radar Library, London, UK).

Schneiderman, H. and Kanade, T. (2004). Object detection using the statistics of parts, *International Journal of Computer Vision* **56**, 3, pp. 151–177.

Shapiro, L. and Stockman, G. (2001). *Computer Vision* (Prentice Hall, Upper Saddle River, NJ, USA).

Skocaj, D., Leonardis, A. and Bischof, H. (2007). Weighted and robust learning of subspace representations, *Pattern Recognition* **40**, 5, pp. 1556–1569.

Spengler, M. and Schiele, B. (2001). Towards robust multi-cue integration for visual tracking, in *Proceedings of the Second International Workshop on Computer Vision Systems (ICVS '01)* (Springer-Verlag, London, UK), pp. 93–106.

Stauffer, C. and Grimson, W. E. L. (2000). Learning patterns of activity using real-time tracking, *IEEE Transactions on Pattern Analysis and Machine Intelligence* **22**, 8, pp. 747–757.

Stergiou, A., Pnevmatikakis, A. and Polymenakos, L. (2007). The AIT outdoor tracker for vehicles and pedestrians in CLEAR 2007, in *CLEAR'07 Evaluation Campaign and Workshop - Classification of Events, Activities and Relationships* (Baltimore, MD, USA), pp. 148–159.

Strobl, K. H., Sepp, W., Fuchs, S., Paredes, C. and Arbter, K. (2005). DLR CalDe and DLR CalLab, http://www.dlr.de/rm/desktopdefault.aspx/tabid-3925/6084_read-9201/.

Swain, M. J. and Ballard, D. H. (1991). Color indexing, *International Journal of Computer Vision* **7**, 1, pp. 11–32.

Swokowski, E. (1979). *Calculus with Analytic Geometry* (Prindle, Weber, and Schmidt, Boston, USA).

Talantzis, F. and Constantinides, A. (2009). Using information theory to detect voice activity, in *Proceedings of IEEE International Conference on Acoustics, Speech, and Signal Processing (ICASSP)* (Taipei, Taiwan), pp. 4613–4616.

Talantzis, F., Constantinides, A. G. and Polymenakos, L. C. (2005). Estimation of direction of arrival using information theory, *IEEE Signal Processing Letters* **12**, 8, pp. 561–564.

Talantzis, F., Pnevmatikakis, A. and Constantinides, A. G. (2008). Audio-visual active speaker tracking in cluttered indoors environments, *IEEE Transactions on Systems, Man, and Cybernetics, Part B: Cybernetics* **38**, 3.

Tekalp, A. M. (1995). *Digital video processing* (Prentice Hall, Upper Saddle River, NJ, USA).

Tranter, S. E. and Reynolds, D. A. (2006). An overview of automatic speaker diarization systems, *IEEE Transactions on Audio, Speech and Language Processing* **14**, 5, pp. 1557–1565.

Tyagi, A., Potamianos, G., Davis, J. W. and Chu, S. M. (2007). Fusion of multiple camera views for kernel-based 3D tracking, in *Proceedings of the IEEE Workshop on Motion and Video Computing (WMVC)* (Austin, TX, USA), pp. 1–8.

Veeraraghavan, H., Schrater, P. and Papanikolopoulos, N. (2006). Robust target detection and tracking through integration of motion, color, and geometry, *Computer Vision and Image Understanding* **103**, 2, pp. 121–138.

Vermaak, J. and Blake, A. (2001). Nonlinear filtering for speaker tracking in noisy and reverberant environments, in *IEEE International Conference on Acoustics, Speech and Signal Processing (ICASSP)*, Vol. 5 (Salt Lake City, USA), pp. 3021–3024.

Viola, P. A. and Jones, M. J. (2001). Rapid object detection using a boosted cascade of simple features, in *IEEE Computer Society Conference on Computer Vision and Pattern Recognition (CVPR 2001)* (Kauai, HI, USA), pp. 511–518.

Waibel, A. and Stiefelhagen, R. (eds.) (2009). *Computers in the Human Interaction Loop* (Springer-Verlag, London, UK).

Ward, D., Lehmann, E. and Williamson, R. (2003). Particle filtering algorithms for tracking an acoustic source in a reverberant environment, *IEEE Transactions on Acoustics Speech and Signal Processing* **11**, 6, pp. 826–836.

Welch, G. and Bishop, G. (1995). *An Introduction to the Kalman Filter* (University of North Carolina at Chapel Hill, Chapel Hill, NC, USA).

Weng, S.-K., Kuo, C.-M. and Tu, S.-K. (2006). Video object tracking using adaptive Kalman filter, *Journal of Visual Communication and Image Representation* **17**, 6, pp. 1190–1208.

Wrigley, S., Brown, G., Wan, V. and Renals, S. (2005). Speech and crosstalk detection in multichannel audio, *IEEE Transactions on Speech and Audio Processing* **13**, 1, pp. 84–91.

Yang, M.-H. (2004). Recent advances in face detection, in *IEEE International Conference on Pattern Recognition (ICPR 2004)* (Cambridge, UK).

Zhang, Z. (2000). A flexible new technique for camera calibration, *IEEE Transactions on Pattern Analysis and Machine Intelligence* **22**, 11, pp. 1330–1334.

Zhang, Z., Potamianos, G., Senior, A. W. and Huang, T. S. (2007). Joint face and head tracking inside multi-camera smart rooms, *Signal, Image and Video Processing* **1**, 2, pp. 163–178.

Zhu, Z. and Huang, T. (eds.) (2007). *Multimodal Surveillance: Sensors, Algorithms and Systems* (Artech House, Boston - London).

Zotkin, N., Dmitry, Duraiswami, R. and Davis, S., Larry (2002). Joint audio-visual tracking using particle filters, *EURASIP Journal on Applied Signal Processing* **2002**, 11, pp. 1154–1164.

Index

www.ingramcontent.com/pod-product-compliance
Lightning Source LLC
Chambersburg PA
CBHW050558190326
41458CB00007B/2085

9 781848 165816